THE DIVIDED MIND

Also by John E. Sarno, M.D.:

Mind Over Back Pain: A Radically New Approach to the Diagnosis and Treatment of Back Pain

The Mindbody Prescription: Healing the Body, Healing the Pain

Healing Back Pain: The Mind-Body Connection

Stroke: The Condition and the Patient
coauthored with Martha Taylor Sarno

Stroke: A Guide for Patients and Their Families
coauthored with Martha Taylor Sarno

THE DIVIDED MIND

THE EPIDEMIC OF MINDBODY DISORDERS

MIND
THE DIVIDED

JOHN E. SARNO, M.D.

WITH CONTRIBUTIONS FROM SAMUEL J. MANN, M.D., IRA RASHBAUM, M.D., ANDREA LEONARD-SEGAL, M.D., JAMES R. ROCHELLE, M.D., DOUGLAS HOFFMAN, M.D., AND MARC SOPHER, M.D.

HARPER

NEW YORK • LONDON • TORONTO • SYDNEY

A hardcover edition of this book was published in 2006 by HarperCollins Publishers.

Grateful acknowledgment is made for permission to print the following material:

"Hypertension and the Mindbody Connection: A New Paradigm" © Samuel J. Mann, M.D. Used by permission of the author.

"My Experience with Tension Myositis Syndrome" © Ira Rashbaum, M.D. Used by permission of the author.

"A Rheumatologist's Experience with Psychosomatic Disorders" © Andrea Leonard-Segal, M.D. Used by permission of the author.

"My Perspective on Psychosomatic Medicine" © James R. Rochelle, M.D. Used by permission of the author.

"Structural Pain or Psychosomatic Pain?" © Douglas Hoffman, M.D. Used by permission of the author.

"A Family Doctor's Experience with Mindbody Medicine" © Marc Sopher, M.D. Used by permission of the author.

First paperback edition published 2007.

Designed by Kris Tobiassen

The Library of Congress has catalogued the hardcover edition as follows:

Sarno, John E., 1923–
The divided mind : the epidemic of mindbody disorders / John E. Sarno.
p. cm.
Includes bibliographical references.
ISBN 0-06-085178-3
1. Medicine, Psychosomatic—United States. 2. Mind and body—United States. I. Title.

RC49.S339 2006
616.08—dc22 2005055267

24 25 26 27 28 LBC 35 34 33 32 31

ISBN 13: 978-0-06-117430-8 (pbk.)
ISBN 10: 0-06-117430-0 (pbk.)

CONTENTS

INTRODUCTION

Health care in America is in a state of crisis. Certain segments of American medicine have been transformed into a dysfunctional nightmare of irresponsible practices, dangerous procedures, bureaucratic regulations, and skyrocketing costs. Instead of healing people, the broken health care system is prolonging people's suffering in too many cases. Instead of preventing epidemics, it is generating them.

Does this judgment sound too harsh? Let's look at some statistics. Over six million Americans suffering from the mysterious and excruciatingly painful ailment called "fibromyalgia" are being treated by an army of self-minted specialists, not one of whom has a clue as to what causes the disorder. Millions more are suddenly being treated for gastric reflux, at an annual cost of billions of dollars. Who says heartburn can't be profitable? And millions more—many of them youngsters—are dependent on mind-altering drugs which, it now turns out, may actually be endangering their lives.

The circumstances are serious. I am not overstating the situation. That's why my colleagues and I have written this book.

The Divided Mind is about the principles and practice of psychosomatic medicine. It is not about alternative medicine, or some trendy New Age regimen. It is about straightforward, clinically tested medicine, as practiced by licensed physicians for over thirty years, working with thousands of patients.

First, I want to clear up any confusion surrounding the word *psychosomatic*. You may think it refers to something vaguely fraudulent, such as imaginary diseases dreamed up by people for their own selfish or confused reasons. That's simply not true. But even medical practitioners, doctors who might be expected to have a more accurate understanding of the term, sometimes make the mistake of assuming it refers to how stress makes disease worse, or the stressful consequences of living with a disease. Those are legitimate concerns and have been addressed in the medical literature, but they are not psychosomatic. Psychosomatic medicine specifically refers to physical disorders of the mindbody, disorders that may appear to be purely physical, but which have their origin in unconscious emotions, a very different and extremely important medical matter. Note that we will use the terms *psychosomatic* and *mindbody* interchangeably throughout the book, so don't let it throw you.

There are literally hundreds of disorders and illnesses that have been identified as purely psychosomatic or having a psychosomatic component. We will explore many of them in the pages that follow. They can range from mildly bothersome back pain all the way to cancer, depending on the power and importance of unconscious emotional phenomena. Psychosomatic illnesses seem to be an inescapable part of the human condition. Yet amazingly, in spite of the nearly universal prevalence of such disorders, the practice of psychosomatic medicine is almost totally unknown within today's medical community, and plays virtually no part in contemporary medical study and research. Nowadays, when physicians and many psychiatrists are confronted with a psychosomatic disorder, they do not recognize it for what it is and almost invariably treat the symptom.

The enormity of this miscarriage of medical practice may be compared to what would exist if medicine refused to acknowledge the existence of bacteria and viruses. Perhaps the most heinous manifestation of this scientific medievalism has been the elimination of the term *psychosomatic* from recent editions of the *Diagnostic and Statistical Manual of Mental Disorders* (DSM), the official publication of the American Psychiatric Association. One might as well eliminate the word *infection* from medical dictionaries.

This astonishing state of affairs—scandalous really—did not occur overnight. For the first half of the twentieth century the study and treatment of psychosomatic disorders was recognized by many medical professionals as a promising and important new frontier of medicine. Then, about fifty years ago, the American medical community took a wrong turn and simply abandoned its interest in psychosomatic medicine. I shall speculate on why this happened, but for now the important thing to note is that as a direct result of turning its back on this vital branch of medicine, the medical profession has helped to spawn epidemics of pain and other common disorders affecting the lives of millions of Americans.

I came upon psychosomatic medicine well along in my professional career, when I began to see large numbers of people suffering from those common but sometimes mysterious conditions associated with bodily pain, primarily of the low back, neck, trunk, and limbs. I did not know these disorders were psychosomatic. I had not trained in psychiatry or psychology, and it was only through direct daily confrontation with the suffering of my patients that I eventually came to recognize the true nature of their distress, and could then begin to administer effective treatment. Over the last thirty-two exciting and fruitful years, my colleagues and I have learned much. I've published three books to describe our work, our discoveries, and our successes. Those dealt largely with what I called the tension myositis syndrome

(TMS), a painful psychosomatic disorder afflicting millions. *The Divided Mind* will deal with the full range of psychosomatic disorders, a far broader and more important subject. Psychosomatic disorders fall into two categories:

1. Those disorders that are *directly induced* by unconscious emotions, such as the pain problems (TMS) and common gastrointestinal conditions including reflux, ulcers, irritable bowel syndrome, skin disorders, allergies, and many others.

2. Those diseases in which unconscious emotions may play a role in causation, *but are not the only factor*. They include autoimmune disorders like rheumatoid arthritis, certain cardiovascular conditions, and cancer. No one, as far as I know, who is currently studying these disorders includes unconscious emotions as potential risk factors. To my mind, this borders on the criminal.

Psychosomatic processes begin in the unconscious, that dark, unmapped, and generally misunderstood part of our minds first identified by Sigmund Freud. Though it has yet to be appreciated by either physical or psychiatric medicine, unconscious emotions are a potent factor in virtually all physical, nontraumatic ills. I gave this book the title *The Divided Mind* because it is in the interaction of the unconscious and conscious minds that psychosomatic disorders originate. Those traits that reside in the unconscious that we consider the most troublesome, like childishness, dependency, or the capacity for savage behavior, are the products of an old, primitive part of the brain, anatomically deep, just above the brain stem. Evolution has added what is called the neocortex, the new brain, the brain of reason, higher intelligence, communication, and morality. There appears to be an ongoing struggle between these two parts of the brain. Sometimes reason prevails, and at other times the more childish, bestial part of human nature is dominant. This duality is one reason for psychosomatic disorders, as will be demonstrated.

The conclusions found in this book are not based on armchair deductions. They are the result of many years of experience with thousands of patients, and are reinforced by the findings of highly trained psychotherapists. In addition, six pioneering physicians from around the United States who have incorporated psychosomatic principles in their practices and research have also contributed findings based on their own experiences. Our successful treatment of a remarkably high percentage of patients dynamically supports our findings.

The Divided Mind is intended primarily to explain the nature of the psychosomatic process, particularly the psychology that leads to clearly obvious physical symptoms. The book's secondary purpose is to draw attention to the blinkered attitudes of too many practitioners of contemporary medicine who fail not only to acknowledge the existence of psychosomatic disorders, but who actually contribute to their spread by their failure to do so.

I undoubtedly will be challenged by the guardians of perceived wisdom for the so-called "lack of scientific evidence" for my diagnostic theories. This is almost ludicrous since there is no scientific evidence for some of the most cherished conventional concepts of symptom causation. The most glaring example of this is the idea that an inflammatory process is responsible for many painful states, for which there is no scientific evidence. Another example: studies have never been done to validate the value of a variety of surgical procedures employed for pain disorders, like laminectomy for intervertebral disk abnormalities.

Studying psychosomatic disorders in the laboratory poses some great problems. How do you identify and measure unconscious emotions? If acceptance of the diagnosis by the patient is critical to successful treatment, how can you demonstrate the validity of the diagnosis and treatment if most of the population doesn't accept the diagnosis? After many years of experience it is our impression that not more then 10 to 15 percent of the population would be willing to accept a psychosomatic diagnosis. Our proof of validity is the remarkable success of our therapeutic program.

As Freud noted, the physiology of the process is far less important than accurate observations of the process itself. He didn't have any laboratory data either. So I must leave it to the laboratory experts to figure out the nuts and bolts of the process.

By sharing with you focused experiences in the diagnosis and treatment of large numbers of people who have suffered and are suffering from psychosomatic pain, my fellow doctors and I hope that our findings will have an important influence on medical practice, particularly in view of the millions who now suffer these disorders needlessly.

In conclusion I must express my deep gratitude to Mr. Al Zuckerman, who succeeded in finding a publisher for this rather controversial book.

ONE

WHAT IS PSYCHOSOMATIC MEDICINE?

I remember the first time John R came into my clinic in 1996. He was a successful businessman in his early forties, well dressed and fit, radiating confidence. He seemed altogether at ease and self-assured—until he bent to sit down. Abruptly, his movements slowed and he became so cautious, so fragile, so tentative that he was suddenly a caricature of the driving, confident man who strode through my door only moments before. His body language made it clear that he was either experiencing excruciating pain or feared the pain would strike him if he made the slightest wrong move.

As a medical doctor, I could empathize with his suffering. My specialty is mindbody disorders, and I see cases like this every working day. I hoped I could help him, which meant helping him to help himself, because with mindbody disorders, a doctor cannot "cure" a patient. It is the suffering patient who must come to understand his malady . . . and by understanding it, banish it.

As we went over John R's history, a picture began to emerge of an interesting and satisfying life. Married, three children. His own business, which probably took up too much of his time, but was doing well. I also heard a familiar litany of suffering and pain—a chronic bad back of mysterious origins, sometimes inducing such severe pain that he could not get out of bed in the morning. His long and unsuccessful search for relief—experiments with alternative medicine, prescription drugs, and finally, in desperation, surgery—immensely expensive and only temporarily successful. Then the sudden onset of brand-new ailments: sciatica, migraine headaches, acid reflux—the list of maladies went on and on.

As a physician, my heart went out to him. It was my job to help him. But I could only lead. Would John R follow? Would he understand the profound interconnectedness of mind and body? Would he grasp the awesome power of buried rage?

To the uninitiated, there is often something mysterious about mindbody medicine. In truth, the relationship of the mind to the body is no more mysterious than the relationship of the heart to the circulation of the blood, or that of any other organ to the workings of the human body. My first interview with John R indicated he would be open to the idea of mindbody medicine. Within a month of beginning treatment, his pains, which had tortured him for much of his adult life, simply disappeared, without the use of drugs or radical procedures. I still get an annual Christmas card from him. In his most recent one he reported that he continues playing tennis and skiing. Last summer he and his oldest boy walked the entire Appalachian Trail. The pain and the equally unexplained other disorders have not returned.

Many of my patients have an initial difficulty grasping the full dynamics of the mindbody syndrome. It is one thing to accept the concept that the mind has great power over the body, but quite another to internalize that knowledge, and to understand it on a deeply personal basis. Even when my patients come to fully appreciate the central element of the equation—that it is their *mind* that contains the root cause of their physical distress—they may continue to stumble over the secondary details, unable to accept the reality of their own buried rage, and remain puzzled over the fact that their own mind can make decisions of which they are unaware.

Sometimes it helps my patients to understand the mindbody connection if they step back and look at it from a broader perspective. Psychosomatic disorders belong to a larger group of entities known as *psychogenic* disorders, which can be defined as any physical disorders induced or modified by the brain for psychological reasons.

Some of these manifestations are commonplace and familiar to all, such as the act of blushing, or the feeling of butterflies in the stomach, or perspiring when in the spotlight. But these are harmless and temporary phenomena, persisting only as long as the unusual stimulus remains.

A second group of psychogenic disorders includes those cases in which the *pain of a physical disorder is intensified* by anxieties and concerns not directly related to the unusual condition. An example would be someone recently involved in a serious automobile accident whose pain may be significantly worsened by concerns about his or her family, job, and so on, not about the injuries. While mainstream medicine tends to ignore almost all psychogenic manifestations, it generally acknowledges this type, recognizing that symptoms may worsen if the patient is anxious. Doctors may refer to this as *emotional overlay*. In my practice, patients have reported that their pain became much more severe when they were informed of the results of a magnetic resonance imaging (MRI) scan that described an abnormality, such as a herniated disk, particularly if surgery was suggested as a possible treatment.

The third psychogenic group is the exact opposite of the second: it covers cases in which there is a *reduction of physical symptoms* in an existing disorder. In one of the earliest studies of pain, Henry Beecher of Harvard reported that in a group of severely wounded soldiers in World War II, it was found that despite the severity of their injuries they often required little or no analgesic medication because their pain was substantially lessened by their becoming aware that they were still alive, being cared for and removed from the dangers of deprivation, hardship, and sudden death.

By far the most important psychogenic categories are the fourth and fifth groups, *hysterical disorders* and *psychosomatic disorders*. Hysterical disorders are mostly of historical interest, although the psychology of both is identical. My experience has been primarily with psychosomatic disorders.

The symptoms of hysterical disorders are often quite bizarre. The patient may experience a wide variety of highly debilitating maladies, including muscle weakness or paralysis, feelings of numbness or tingling, total absence of sensation, blindness, inability to use their vocal cords, and many others, all *without any physical abnormalities in the body to account for such symptoms*.

It is clear from the nature of hysterical symptoms that their origin is indeed "all in the head," to take a pejorative phrase commonly used to refer to psychosomatic symptoms. The absence of any physical change to the body indicates that the symptoms are generated by powerful emotions in the brain. Just where in the brain, no one can say for sure. One medical authority, Dr. Antonio R. Damasio, has suggested that these emotion-generating centers are located in the hypothalamus, amygdala, basal forebrain, and brain stem. The patients perceive symptoms as though they were originating in the body when the appropriate brain cells are stimulated. These symptoms often have a very strange and unreal quality about them. One of the nineteenth-century pioneers of psychiatry, Josef Breuer, likened them to hallucinations.

PSYCHOSOMATIC DISORDERS

By contrast, in the fifth psychogenic group, psychosomatic disorders, the brain induces *actual physical changes* in the body. An example of this would be tension myositis syndrome (TMS), a painful disorder that we will examine at greater length. In this condition, the brain orders a reduction of blood flow to a specific part of the body, resulting in mild oxygen deprivation, which causes pain and other symptoms, depending on what tissues have been oxygen deprived.

One of the most intriguing aspects of both hysterical and psychosomatic disorders is that they tend to spread through the population in epidemic fashion, almost as if they were bacteriological in nature, which they are not. Edward Shorter, a medical historian, concluded from his study of the medical literature that the incidence of a psychogenic disorder grows to epidemic proportions when the disorder is in vogue. Strange as it may seem, people with an unconscious psychological need for symptoms tend to develop a disorder that is well known, like back pain, hay fever, or eczema. This is not a conscious decision.

A second cause of such epidemics often results when a psychosomatic disorder is misread by the medical profession and is attributed to a structural abnormality, such as a bone spur, herniated disc, etc.

A 1996 study in Norway suggests there is a third condition that fuels such epidemics: the simple fact that medical treatment may be readily available. A paper published in the journal *Lancet* in 1996 described an epidemic in Norway of what is called "whiplash syndrome." People involved in rear-end collisions, though not seriously injured, were developing pain in the neck and shoulders following the incident. Norwegian doctors were puzzled by the epidemic and decided to investigate. They went to Lithuania, a country with no medical insurance, and on the basis of a controlled study determined that the whiplash syndrome simply did not exist in that country. It turned out that the prevalence of whiplash in Norway had less to do with the severity of rear-end collisions than with the fact that it was in vogue; doctors couldn't explain the epidemic and the ready availability of good medical insurance for treatment!

The most important epidemics of psychosomatic disorders are those associated with pain. As will be discussed below, they have become the ailments du jour for millions of Americans. They are "popular" and most of them have been misdiagnosed as being the result of a variety of physical structural abnormalities, hence their spread in epidemic fashion.

What is the genesis of a psychosomatic disorder? As we shall see, the cause is to be found in the unconscious regions of the mind, and as we shall also see, its purpose is to deliberately distract the conscious mind.

The type of symptom and its location in the body is not important so long as it fulfills its purpose of diverting attention from what is transpiring in the unconscious. On occasion, however, the choice of symptom location may even contribute to the diversion process, something that is common with psychosomatic disorders. For example, a man who experiences the acute onset of pain in his arm while swinging a tennis racket will naturally assume that it was something about the swing that hurt his arm. The reality is that his brain has decided that the time is ripe for a physical diversion and chooses that moment to initiate the pain, because the person will assume that it stems from an injury, not a brain-generated physical condition that caused the pain. How does the brain manage this trick? It simply renders a tendon in the arm slightly oxygen deprived, which results in pain. This is how "tennis elbow" got its name. If that sounds bizarre, diabolical, or self-destructive, you will see later that it is in reality a protective maneuver. My colleagues and I have observed it in thousands of patients.

But in time, such a symptom may lose its power to distract. Then the psyche has another trick up its sleeve. It will find another symptom to take its place, one that is viewed by both patient and doctor as "physical," that is, not psychological in origin. For instance, if a treatment—let's say surgery—neutralizes a particular psychogenic symptom, so that the symptom loses its power to distract, the brain will simply find another target and create another set of symptoms. I have called this the *symptom imperative* and it has enormous public health implications, because psychogenic symptoms are commonly misinterpreted and treated as physical disorders. All of a sudden, the "cured" patient has a brand-new disorder that demands medical attention. More distress. More time lost. More expense. This will be documented as we proceed.

Statistically, the most common psychosomatic disorder today is TMS, which I have described in its many forms in my previous books. I gave it that name because at the time of publication of the first book in 1984, it was thought that muscle (myo) was the only tissue involved. Since then, I have come to learn that nerve and tendon tissue may also be targeted by the brain; in fact, it now appears that nerve involvement is more common than muscle. Accordingly, a more inclusive name, like *musculoskeletal mindbody syndrome*, might be more appropriate. However, because the term *TMS* is now so well known, I have been urged by my colleagues not to change it, so TMS it remains.

DISORDERS MEDIATED THROUGH THE AUTONOMIC-PEPTIDE SYSTEM

How does the brain induce symptoms in the body? There are a number of ways, but by far the largest number of psychosomatic conditions are created through the activity of the *autonomic-peptide system*. The autonomic branch of the central nervous system controls the involuntary systems in the body, such as the circulatory, gastrointestinal, and genitourinary systems. It is active twenty-four hours a day and functions outside of our awareness. The word *peptides* has been added because peptides are molecules that participate in a system of intercommunication between the brain and the body and play an important part in these processes.

The most common disorders produced through this system are those of TMS, described above. These disorders afflict millions and cost the economy billions of dollars every year in medical expenses, lost work time, compensation payments, and the like.

Other conditions include:

- Gastroesophageal reflux
- Peptic ulcer (often aggravated by anti-inflammatory drugs)
- Esophagospasm
- Hiatus hernia
- Irritable bowel syndrome
- Spastic colitis

- Tension headache
- Migraine headache
- Frequent urination (when not related to medical conditions such as diabetes)
- Most cases of prostatitis and sexual dysfunction
- Tinnitus (ringing in the ears) or dizziness not related to neurological disease

The theories advanced here are based almost exclusively on work done with TMS, but there are many less common mindbody disorders (like reflux) whose symptoms are also created by the autonomic-peptide system. We refer to these as equivalents of TMS since they are the result of the same psychological conditions that are responsible for TMS. What put me onto the possibility that the pain I was seeing in the early 1970s was psychosomatic was the fact that so many of the pain patients had experienced these equivalent disorders, all of which I knew to be psychosomatic. That realization suggested that the pain disorder I was seeing was also psychosomatic.

WHY TMS IS PAINFUL

As I stated earlier, the altered physiology in TMS appears to be a mild, localized reduction in blood flow to a small region or a specific body structure, such as a spinal nerve, resulting in a state of mild oxygen deprivation. The result is pain, the primary symptom of TMS. The tissues that may be targeted by the brain include the muscles of the neck, shoulders, back, or buttocks; any spinal or peripheral nerve; and any tendon. As a consequence, symptoms may occur virtually anywhere in the body. The nature of the pain varies depending on the tissues involved: muscle, nerve, or tendon. In addition to pain, nerve involvement brings with it the possibility of feelings of numbness and tingling and/or actual muscle weakness. These reflect the function of nerves, which is to bring sensory information to the brain and carry movement messages to the body, either or both of which may be affected in TMS. The fact that patients recover rapidly when they are appropriately treated suggests that the tissues involved—nerve tissue being the most sensitive—are not in any way damaged but only rendered temporarily dysfunctional.

Because so few members of the medical profession recognize mindbody disorders for what they are, the pain of TMS is commonly attributed to a structural abnormality, such as the ones that often show up on x-rays, computed tomography (CT), or MRI scans. Following is a list of the most common ones:

Abnormalities of the intervertebral disc due to wear and tear, aging, etc., including:

- Narrowing of the disc space, indicating that the disc has lost substance
- Bulging of the disc, due to pressure from the material inside the disc (the nucleus pulposus)
- Herniation of disc material

Abnormalities of other spinal bone elements, referred to as spondylosis (immobility and fusion of vertebral joints) including:

- Bone spurs around spinal bone joints ("pinched nerve")
- Enlargement of ligaments in the spinal canal
- Narrowing of the spinal canal due to the changes above (spinal stenosis)
- Spondylolisthesis (malalignment of spinal bones)

- Scoliosis (an abnormal side-to-side curvature of the spine)
- Abnormalities of tendons of rotator cuff muscles in the shoulder
- Tears of the knee cartilage (meniscus)
- Normal aging changes in the knee, called arthritis
- Changes in the hip caused by aging changes (arthritis)
- Bone spurs in the heel of the foot
- Many others less common conditions

In my experience, the majority of these abnormalities are not responsible for the pain. The cause of the pain is TMS, plain and simple. Nevertheless, despite the absence of proof that the abnormalities are the cause of the pain, the medical profession routinely treats those with surgery—in many cases, exorbitantly expensive surgery—as will be detailed.

To further complicate the problem, there are a number of soft tissue disorders that are also blamed for the pain of TMS. These misdiagnoses include:

- Myofascial pain, usually in the back (actual cause unknown)
- The postpolio syndrome (pain in parts of the body previously afflicted by polio). Such pain is routinely attributed to the polio, but there is no proof that this is the cause. There is a Latin phrase commonly quoted in scientific circles that refers to this particular kind of misdiagnosis: "post hoc ergo propter hoc." It means "after this [i.e, polio] therefore because of this," a classic error in logic leading to a dangerous and unscientific conclusion.
- Strained back or neck muscles

- Pain in the buttock attributed to compression of the sciatic nerve by the piriformis muscle—a rather frivolous concept with no evidence of validity
- Pain and other dental abnormalities (temporomandibular joint disorder [TMS]) that are most likely due to TMS in jaw muscles
- Tendon pain in various locations around the elbow attributed to overuse (tennis elbow)
- Wear or tear of rotator cuff tendons
- Pain in the front of the sole of the foot (metatarsalgia)
- Pain in the middle of the sole of the foot (plantar fasciitis)
- Pain in the heel of the foot (bone spur)
- Pain attributed to a benign tumor in the sole of the foot (metatarsal neuroma)
- Carpal tunnel syndrome (repetitive stress injury)
- Fibromyalgia: see what follows
- Other less common soft tissue disorders

In the last thirty-five years, three of the above conditions have been so often misdiagnosed that their incidence has reached epidemic proportions. They are:

1. Chronic pain from back, neck, shoulder, and limbs

2. Fibromyalgia

3. Carpal tunnel syndrome

Each has a different story to tell.

THREE MINDBODY EPIDEMICS

1. Chronic Pain Syndromes: A Modern Plague

The so-called black plague of European and Asian history—bubonic plague—killed millions. It was caused by a bacterium carried by rats and transmitted by fleas. The authorities of the day had the means to control the spread of the plague, but because bacteriology and epidemiology were unknown sciences at that time, they did not understand the need to do so. In other words, the plague flourished because of their ignorance. An epidemic of chronic pain exists today because of a similar lack of knowledge. Modern medicine knows neither the cause of chronic pain nor the means of its spread. This has led directly to an epidemic that has been going on since the late 1960s. It reached its peak in the 1990s and is still devastating the lives of millions. It is why pain clinics have proliferated in recent years.

The reason for this epidemic is the stubborn resistance of the medical profession to even consider the likelihood of mindbody disorders. Most people with chronic pain are suffering from one of the many manifestations of TMS just described, but the majority of practitioners called upon to treat them are unaware of that diagnosis. Those few who know about it often choose not to acknowledge it. Instead, they attribute the pain to one of the many disorders just listed. The persistence of the pain—the fact that it often lasts for months or even years—is explained by an ingenious idea conceived by behavioral psychologists many years ago. According to their theory, the pain continues because it serves the purpose of what is called *secondary gain*, that is, an unconscious desire on the part of the sufferer for some kind of benefit from the symptom, such as sympathy, support, release from responsibility or from arduous labor, monetary gain, and so on. This clever explanation was readily embraced by medical practitioners since it absolved them of responsibility for their failure to help their patients. It was, after all, the patient's own fault. One cannot imagine a more devastatingly wrong explanation, from both the scientific perspective and that of the suffering patient.

As we shall see, the true cause of the pain, TMS, serves the purpose of *primary gain*, that is, to prevent the conscious brain from becoming aware of unconscious feelings like rage or emotional pain. There is rarely secondary gain. We shall elaborate on this in the chapter on the psychology of these disorders.

As noted above, mindbody disorders tend to spread in epidemic fashion:

a. if they are in vogue;
b. if they are misdiagnosed, that is, if the pain is falsely attributed to some purely "physical" phenomenon, like a herniated disc or bacteria in the stomach; and
c. if treatment is readily available and funded by medical insurance.

Chronic pain fits these criteria admirably, which explains the persistent inability of medicine to make any inroads on the problem. The medical profession bears a heavy responsibility for this and for the other epidemics. On the simplest level, it has violated one of its most fundamental medical admonitions: *do no harm*.

In truth, American medicine has done enormous harm. It has misdiagnosed the cause of the pain, guaranteeing that even if the patient experiences pain relief due to a placebo reaction, the pain will return to the same or some other location or, following the principle of the symptom imperative, another physical disorder will take its place. In no way has the patient been healed.

In its blindness, modern medicine has enhanced the tendency of the pain syndromes to spread in epidemic fashion. It has introduced a variety of ineffective treatments, some of them extremely expensive, placing great burdens on the government and private insurance.

The enormity of the problem is illustrated by an article that appeared in the business section of the *New York Times* on December 31, 2003. It described how one such expensive treatment, spinal fusion, is being widely performed *despite the lack of evidence that it has any value whatsoever*. The article went on to point out that the doctors, hospitals, and manufacturers of the hardware used in these procedures all have a financial stake in the performance of this operation. The national bill for its *hardware alone* has soared to $2.5 billion a year. What the cost of treatment must be staggers the imagination. My medical school professors would be shocked and horrified at what has happened to medical practice. The marketplace and economic factors have taken over.

In my experience, the many structural abnormalities that are claimed to be the basis for the surgery described above are usually not responsible for the pain so that neither surgical nor even conservative physical treatment of any kind is appropriate. I have taken to advising my patients that the worst indication for musculoskeletal surgery is pain attributed to some structural abnormality.

2. Fibromyalgia

Fibromyalgia is a medical term that has been around for a long time. For some reason it was adopted by the rheumatology community in the early 1980s and applied to patients suffering pain in many locations in the trunk, arms, and legs. In fact, it is a severe form of TMS. Significantly, fibromyalgia patients commonly suffer from other mindbody disorders as well, like headache and irritable bowel syndrome, as well as emotional symptoms including anxiety, depression, and sleep disorders. When rheumatologists first became interested in people with these symptoms, they were not able to explain what caused the disorder, but they created diagnostic criteria to define it. That became a kind of medical kiss of death. The American College of Rheumatology decreed that the diagnosis could be made if the person under examination exhibited pain in eleven of a potential eighteen locations. Since that time, hundreds, if not thousands, of papers have been published describing studies that try, still unsuccessfully, to explain the disorder. Two of these published studies of people with fibromyalgia found that the oxygen level in their muscles was reduced, confirming the hypothesis that fibromyalgia is a manifestation of TMS, which as we've seen is caused by mild oxygen deprivation. But the rheumatology community did not accept the idea of mild oxygen deprivation as the cause of fibromyalgia, and the epidemic continued. By the year 2000 the enormous increase in the number of people with this diagnosis prompted an article in *The New Yorker* magazine by Jerome Groopman, a professor of medicine at Harvard, in which he noted that there were *six million* Americans (mostly women) with this disorder of unknown cause and that it appeared to be analogous to the nineteenth-century epidemic of neurasthenia.

The fibromyalgia story is another tragic example of the epidemic proclivity of psychosomatic disorders when they are misdiagnosed and, therefore, inevitably mismanaged. Another major epidemic began around the same time, and for the same reason.

3. Carpal Tunnel Syndrome

Carpal tunnel syndrome (CTS) became fashionable in the 1980s. It is another TMS manifestation that has been widely misread by medicine, with predictable results. Patients experience a variety of symptoms in their hands that are the result of dysfunction of the median nerve at the wrist. The dysfunction can be documented by electrical tests, so there is no doubt about the reason for the symptoms. What is in doubt (although the medical community does not admit to any doubt) is what is troubling the nerve. The generally accepted diagnosis is that the nerve is compressed as it passes under a ligament at the wrist, and the recommended treatment is to inject a steroid under the ligament, or cut it, which sometimes produces symptomatic relief. However, a paper published in the journal *Muscle and Nerve* suggests that nerve function returns too rapidly after the ligament has been cut to blame compression for the disorder, and that it is more likely that local ischemia (reduced blood flow) is responsible for the symptoms. Because ischemia is what causes the symptoms of TMS, the finding supports the idea that carpal tunnel syndrome is a manifestation of TMS.

It is highly significant that the rapid spread of carpal tunnel syndrome coincided with the spectacular growth of the computer industry. What fueled the spread of CTS was the belief that the problem was caused by working at computer keyboards, and that CTS was one of a number of "repetitive stress injuries." Since those early days, armies of office workers and those employed in other occupations requiring a variety of repetitive tasks have developed CTS, so that now, like chronic pain and so-called fibromyalgia, it is a major public health problem. People with CTS are particularly resistant to the idea that it is a *mindbody disorder* even when that more benign term is used rather than the word *psychosomatic.*

It is quite remarkable that I have been unable to find a single mention in the medical literature questioning the reason for these epidemics. And one never gets a reasonable answer when one asks, *Why is it that the millions of men and women who pounded typewriters since the beginning of the twentieth century never developed CTS?* Again, medicine bears the responsibility for these epidemics on two counts: first, by failing to make the correct diagnosis, and then by attributing the epidemics to structural and other specious causes, thereby contributing to the severity and long-term nature of symptoms. This is important because it supports the mind's strategy, which is to distract attention from what is going on in the unconscious mind and focus it on a body symptom. By so doing it perpetuates the process. The sad reality is that most of the people suffering from conditions like chronic pain, fibromyalgia, and CTS will not accept a psychosomatic diagnosis.

OTHER DISORDERS MEDIATED THROUGH THE AUTONOMIC-PEPTIDE SYSTEM

In addition to the three conditions just discussed, there are numerous other disorders that are brought about by the same body-mind system, the autonomic-peptide system. Like the first three, they have the same genesis and serve the same psychological purpose. They include:

Gastrointestinal Mindbody Syndromes

Upper and lower gastrointestinal symptoms continue to be common psychosomatic manifestations. They are treated with a variety of medications, often with success which, as has been noted, is a

Pyrrhic victory since the brain will simply find another place to create psychosomatic symptoms.

Many physicians, including psychiatrists, now refuse to believe that ulcers are psychosomatic, because of the discovery of a bacterium in the stomachs of people with peptic ulcers. It is claimed that patients are cured with antibiotics. This is one of the many examples of medicine's inability to confront the reality of psychosomatosis. The presence of bacteria in the stomachs of some patients in our view is merely part of the process.

Similarly, a paper in the *American Journal of Gastroenterology* attributed irritable bowel syndrome to the presence of bacteria in the colon. Such a conclusion would be ludicrous if it were not tragic, for if this idea gains acceptance among physicians and their patients, this is another psychosomatic disorder whose true cause will be ignored in favor of treating the symptom.

Tension Headache and Migraine Headache

There is no laboratory proof that tension headaches and migraine headaches are psychosomatic, but the clinical experience of treating them as such is impressive. As early as the 1930s and 1940s leading medical authorities published numerous papers on the *psychological* basis for migraine, and all noted that migraines were related to repressed rage. In *Psychosomatic Medicine* (1950), Franz Alexander noted, "The most striking observation is the sudden termination of the attack almost from one minute to another after the patient becomes conscious of his hitherto repressed rage and gives expression to it in abusive words." Note Alexander's reference to *rage*. As will be seen, rage in the unconscious mind is central to understanding virtually all psychosomatic reactions.

The groundbreaking work of Alexander and his colleagues (see chapter 2) has been forgotten. The patients who come into our clinic report that the treatment they have previously received for their migraine or tension headaches is invariably with medications, another example of the regression of contemporary medicine.

Genitourinary Mindbody Syndromes

The perceived need for frequent urination is psychosomatic except when it is related to diabetes; renal, cardiac, or adrenal disease; bladder infection; or an enlarged prostate. It is very common. A careful history will reveal that in many cases the habit of getting up frequently during the night to urinate is not brought on by a full bladder but by a mild form of insomnia. The person is programmed by the unconscious to awaken and then programmed to have the urge to urinate.

It has been documented in the medical literature that prostatitis in young men is commonly related to stress when not due to an obvious infection.

Most sexual dysfunction is psychologically based at any age. Though it is well known that libido decreases with age, emotional factors may still be responsible for sexual difficulties in the elderly.

Tinnitus and Vertigo

Both of these conditions may be signs of disorders of the nerves or the ears, but they are most commonly benign and psychosomatic. I once experienced vertigo that lasted only a few hours. It ended when I was able to identify the psychological basis for it.

DISORDERS ATTRIBUTED TO ACTIVITY OF THE IMMUNE-PEPTIDE SYSTEM

The disorders described up to this point are the most common of a very large group, all of them activated by the autonomic-peptide system. A second group of ailments is associated with the body's *immune system.* (Again, we include the peptide communication system because of the role it plays in the interaction between brain and body.) It is not known what determines the unconscious mind's choice of which system or symptom to employ, but it makes no difference since the purpose of all symptoms is the same—to distract the conscious mind.

With the immune-peptide system, the disorder may be induced by either overactive or underactive immune function. Overactive immune activity leads to:

- Allergic phenomena (e.g., allergic rhinitis, conjunctivitis, sinusitis, asthma)
- A large number of skin problems (e.g., eczema, hives, angioedema, acne, psoriasis)

The question invariably arises, "Aren't allergic reactions caused by allergens, like grass pollens?" The answer is yes, but such allergens are merely *triggers.* They are foreign substances, and the immune system is designed to repel foreign invaders. However, not everyone reacts to grass pollens. If your unconscious mind causes your immune system to overreact, the system is said to be *hyperactive* or *hypersensitive.* Both terms denote an allergic reaction. This excessive sensitivity of your immune system is not to protect you from foreign substances, but to keep your conscious attention focused on the body.

Conversely, the unconscious mind may do the opposite to deflect attention from itself. It may *decrease* the efficiency of the immune system and render the person susceptible to infection. Recurrent infections of any kind are usually an indication of this process. The infections must be treated "medically," but they will continue to recur if they are not treated psychologically as well. It is highly significant that many of the people suffering from the pain of TMS who have participated in our therapeutic program have reported the disappearance of allergies or frequent infections simultaneously with the cessation of pain.

Most people with TMS have a history of one or more of these autonomic or immune system conditions. Indeed, it would be most unusual to find someone who has never experienced one or more mindbody symptoms. One is forced to the conclusion that psychosomatic reactions and, therefore, the emotions that cause them, are universal. It is important to recognize that they are not illnesses; *they are a part of life, part of the human condition.* This should become clear when the psychology of mindbody disorders is described in detail in chapter 3.

DISORDERS PRODUCED BY THE NEUROENDOCRINE-PEPTIDE SYSTEM

There is still a third medium for transferring mindbody disorders from the mind to the body. It is the neuroendocrine-peptide system, which governs the body's hormonal distribution. The disorders associated with it are a small but distinct group of conditions that seem to fall somewhere between the physical and psychological in their manifestations:

- Bulimia
- Anorexia nervosa
- Neurasthenia (known today as chronic fatigue)

The desire to overeat or the inability to eat at all seems to point to some strong emotional factor, though it would not be surprising in today's medical atmosphere for someone to come forward with a purely physical explanation for them. Bulimia and anorexia nervosa are generally treated psychiatrically.

As for neurasthenia, a group of physicians representing three of Britain's royal colleges studied the problem and issued a report in 1996 suggesting that psychological factors were primary in the disorder and that a therapeutic program consisting of physical activity and psychotherapy was the most effective of those tried. There is anecdotal evidence based on numerous letters I have received from readers that exposure to my book, *Healing Back Pain*, has relieved many people with neurasthenia. This is logical since the underlying psychology for this is the same as for TMS.

A paper published in *The New England Journal of Medicine* in 1993 entitled "Neuroendocrine-Immune Interactions" concluded with this statement: "Central nervous system influences on the immune system are well documented and provide a mechanism by which emotional states could influence the course of diseases involving immune function. Whether emotional factors can influence the course of autoimmune disease, cancer and infections in humans is a subject of intense research that has not been satisfactorily resolved at this time."

This paper addressed the influence of the neuroendocrine network on the immune system and so has relevance to the allergic and infectious processes referred to above as well as the broad fields of autoimmune disease and cancer. It is introduced here because it is likely that the neuroendocrine network is also responsible for bulimia, anorexia nervosa, and neurasthenia. Once again, the peptide network provides the mechanism by which emotional states are able to induce physical ones.

Bulimia, anorexia nervosa, and neurasthenia are quasi-physical equivalents of TMS. Experience strongly suggests that *anxiety, depression, and obsessive-compulsive disorder (OCD)*, all purely emotional conditions, are equivalents as well.

Recalling the symptom imperative mentioned earlier, I have observed that some patients, upon being relieved of the pain of TMS by some chemical therapy or a placebo, become anxious or depressed rather than developing another physical symptom. But then when their emotional symptoms were relieved by a tranquilizing or antidepressant medication, their body pains returned! Others who were suffering the symptoms of TMS and OCD simultaneously had relief of both while participating in the TMS therapeutic program.

The conclusion is inescapable that the psychology behind both the physical and affective (emotional) disorders is the same and that people whose pain is replaced by anxiety or depression are also experiencing the symptom imperative. This is a daring statement, for it presumes to express an opinion about the origin of anxiety and depression, disorders in the domain of psychology and psychiatry. Nevertheless, it is being suggested that like psychosomatic symptoms, affective states are also reactions to powerful emotions in the unconscious mind that are threatening to become conscious, and it follows that good medicine requires first acknowledging those unconscious emotions and then dealing with them. Treating anxiety or depression with medications without in-depth psychotherapy is poor medicine, and may even be dangerous if the symptom imperative leads to a serious disorder like one of the many autoimmune maladies or cancers. These are not fanciful conclusions based on conjecture; they derive from irrefutable clinical experience.

A word about the peptide network: the scientist who has contributed most to an understanding of this crucially important system, who has, in fact, written about "the biochemistry of emotions," is Dr. Candace Pert. She has described her work in her book, *Molecules of Emotions*, which should be read by all professionals interested in the mechanics of how emotions induce physical symptoms. The peptide network explains the physical part of the psychosomatic process, but it also explains the placebo effect, namely, how blind faith can lead to the amelioration of symptoms. It has been stated already that the placebo effect may be dangerous because of the symptom imperative, but treating the symptom rather than the cause is poor medicine in any event because it is almost invariably temporary, whether or not it leads to a substitute symptom. Placebos take many forms: surgery, a variety of other physical treatments, and pharmaceuticals. If the celestial architect were to suddenly abolish the placebo effect in humans, there would be economic chaos, particularly in the United States, for much medical treatment today owes its success, such as it is, to the placebo phenomenon.

THE CURRENT STATUS OF MINDBODY MEDICINE

In view of our success in the treatment of pain disorders, my colleagues and I are often asked why more patients and physicians don't subscribe to these theories. It's a good question, and not easy to answer. The reasons are many and some of them subtle.

WHY MOST PEOPLE CANNOT ACCEPT THE IDEA OF MINDBODY DISORDERS

Experience suggests that in the United States only 10 to 20 percent of people with a psychosomatic disorder are able to accept the fact that their symptoms are emotional in origin. Many are downright hostile to the idea. Though there are large numbers who seek psychotherapy or psychoanalysis, they represent only a small portion of the entire population. For the majority there is a stigma attached to disorders relating to psychology. Negative words like *weird, crazy, kooky*, and *nuts* come to mind. Psychologists and psychiatrists are head shrinkers or "shrinks." "It's all in your mind" is almost insulting, implying there's something strange or weak about you or that the symptoms are in your imagination. This is most unfortunate, since the symptoms are very real, the result of a very physical process.

Another factor negatively impacting mindbody medicine is that, as with the stigma attached unfairly to cancer and tuberculosis patients in the early twentieth century, there is shame associated with the idea that one may be suffering from psychologically induced symptoms. This persists in many quarters despite the fact that today's young, educated people are more accepting of the need for psychological help than were earlier generations.

Stress is another matter. Most people will accept the idea of stress, finding it less threatening because they think of stress as stemming from things "out there" that are doing something to you, so it does not imply some personal defect. Much of the research in psychology today has to do with the effects of stress in both health and illness. For example, how does stress make a medical condition like diabetes worse? Or how does a medical condition like diabetes cause stress in one's life? This is laudable research, but it doesn't deal with that crucial domain—the unconscious, which is where mindbody disorders begin.

Much of the skepticism of psychosomatic therapy demonstrated by patients is strongly reinforced by the medical profession, including much of the psychiatry community. People much prefer a diagnosis that suggests they can get better with a "quick fix": an injection, a medication, a manipulation, even surgery. Many patients come to see me only after they have tried all of the above.

WHY THE MEDICAL PROFESSION IGNORES MINDBODY CONCEPTS

Since the mid–twentieth century the physical specialties of medicine have moved increasingly farther away from the idea that the brain can bring about physical alterations in the body and that psychosomatic disorders exist. Some specialties, such as orthopedics, neurosurgery, neurology, and physiatry, are particularly opposed to the idea, no doubt because it contradicts their belief that structural abnormalities account for all observed symptoms. Their diagnoses are based on the therapeutic methods they employ. They are, therefore, understandably loath to consider another diagnosis, particularly one that is psychosomatic. Primary care physicians, who generally do not consider themselves competent to deal with patients suffering from persistent pain or neurological symptoms, tend to refer them to "specialists"—the very orthopedists, neurologists, and the like, who have already rejected the validity of psychosomatic diagnoses. Those same primary care physicians might well choose to treat the disorders themselves, if they understood that they were psychosomatic.

Psychosomatic symptoms involving other systems (e.g., gastrointestinal, genitourinary, dermatologic) are usually treated with medication, diet, and so on. Doctors of all kinds now appear to be constitutionally incapable of attributing physical symptoms to emotions. This is a dramatic change from medical attitudes and practices in the first half of the twentieth century. The legendary Sir William Osler once remarked that one was more likely to learn about the course of tuberculosis by looking into the patient's head than in his chest. What has happened?

First, a sad paradox. Medical research has become more laboratory oriented in the last fifty years. To be sure, this shift has produced some impressive results. But at the same time, human biology is not exclusively mechanical, and there are limits to what the laboratory can accurately study. The laboratory study of infectious diseases has been magnificent—it is very straightforward. But its very success has deflected attention from the influence of emotions. As a result, medical research has failed abysmally in many areas. The evidence is everywhere you look. Pain problems have become epidemic. Gastrointestinal, dermatologic, and allergic conditions are increasingly widespread, all because laboratory identification of the physics and chemistry of these conditions does not, contrary to popular medical belief, identify their cause. And paradoxically, wonderful new diagnostic tools, like the MRI, often contribute to misdiagnoses when doctors misinterpret the significance of findings. The methods of the laboratory may be impeccable but are wasted if the interpretation of their findings is faulty.

The failure of scientific medicine to stem the tide of chronic pain disorders is unfortunate enough, but it has failed as well in another even more crucial sphere. There is abundant anecdotal evidence in the medical literature that psychological factors influence more serious disorders like those of the autoimmune group, cardiovascular conditions, and cancer. Yet scientific medicine has paid scant attention to this evidence in its research, with the National Institutes of Health conspicuous in its indifference. Put bluntly, emotional factors should be studied as risk factors in these life-threatening disorders, and they are not.

Another trend in contemporary scientific medicine is its preoccupation with studying the anatomy, physiology, and chemistry of the brain, at the expense of studying its dynamic relationship to the body as a whole. Neuroscience can be enormously important and of consuming interest, but what is learned of the physical brain may be either detrimental or irrelevant to clinical medicine. An example of the former is the almost universal tendency to treat the chemical aberrations associated with depression with drugs, as though the altered chemistry was the cause of the depression when, in fact, the reason for the depression is an unconscious psychological conflict and the chemical change is merely the mechanism that produces the symptom of depression. Treating the depression with drugs alone, without psychotherapy, is not only poor medicine, it is also dangerous. The symptom imperative tells us that taking away a symptom by the use of a placebo or an antidepressant will only give rise to another symptom, and the other symptom may be related to something serious, like cancer.

Then, too, the findings of neuroscience may be totally irrelevant to some areas of clinical medicine. For example, the fact that a positron-emission tomography (PET) scan can identify the areas of the brain that are activated when a person is manifesting anger is not helpful in determining the source of the anger, particularly if unconscious processes are involved. Such findings are extremely interesting but of little use if one is trying to help a patient deal with a behavioral problem. Such help can come only from the laborious process of psychological analysis conducted by someone appropriately trained. When I am working with a patient suffering from pain induced by buried rage, it does no good to know which brain nuclei are involved in the pain process. I must help the patient to understand the sources of the rage. Experience has demonstrated that such understanding will usually "cure" him. This very interesting and germane process will be explained in chapter 4.

Neuroscience is one of the contemporary glamour specialties of research medicine, thanks to some extent to the interest of people like Drs. Gerald M. Edelman and the late Francis Crick, both Nobelists in other fields. Their studies of the "neural correlates of consciousness" are of enormous interest, comparable to the fascinating work being done by cosmologists and astrophysicists, but of little relevance to clinical medicine, particularly where emotions are involved.

An article in the May 2004 issue of *Natural History* illustrates beautifully the limitations of laboratory findings. The author, Robert M. Sapolsky, a professor of biological sciences and neurology, reported on what he identified as a landmark paper published in the journal *Science*. The investigators followed a population of over a thousand New Zealand children from age three into young adulthood, identifying the incidence of depression, and noting that a proportion of the group being studied also possessed a serotonin-regulating gene known as 5-HTT. The role of serotonin in depression is well known due to widely used drugs like Prozac. The investigators correlated the incidence of two variants of the 5-HTT gene and depression and found that inheriting the genes only increased the risk of depression in people. The "bad" gene did not produce depression in those who had not suffered major stresses. The author noted, "We all have a responsibility to create environments that interact benignly with our genes."

Another aspect of this problem was enunciated by Stephen J. Gould, who wrote in *Natural History*, "An unfortunate but regrettable common stereotype about science divides the profession into two domains of different status. We have, on the one hand, the 'hard' or physical sciences that deal in numerical precision, prediction and experimentation. On the other hand, 'soft' sciences that treat the complex objects of history in all their richness must trade these virtues for 'mere' description without firm numbers in a confusing world where, at best, we can hope to explain what we cannot predict. The history of life embodies all the messiness of this second, and undervalued, style of science."

As this book was being prepared for publication a very important medical paper appeared in the September 2005 issue of the

Proceedings of the National Academy of Sciences. A research team at the University of Wisconsin was able to relate activity in areas of the brain known to be involved with emotions to an inflammatory process that causes symptoms of asthma. Since we theorize that asthma is a mindbody disorder, and an equivalent of TMS, this is important evidence that emotions may be a crucial factor in the causation of mindbody disorders. I intend to initiate a similar study since it is highly likely that the brains of people suffering an episode of TMS will show the same kind of changes.

Neuroscience can play an important role in identifying how mindbody processes work. If unconscious emotions can be identified and measured objectively we would have so-called hard data to support our clinical observations.

The world of the unconscious mind, like the history of life, cannot be studied exclusively by hard science. How can one objectively identify and quantify the personality traits and emotions that reside, so to speak, in the unconscious? The idea that powerful unconscious emotions are responsible for mindbody disorders is based on medical history, knowledge of the psyche, physical examination, logical deduction, and trial-and-error therapeutic experimentation. Success in treatment validates the accuracy of diagnosis if one is assured that there is no placebo effect.

Instead of dealing with this messy reality, contemporary medical science has simply discarded the entire concept of mindbody medicine. It would rather deal with mechanical, measurable, chemical realities than the abstruse phenomena of psychology. It does not want to know that emotions drive the chemical and physical manifestations they have identified, and it has the dangerous idea that treating the chemistry will correct the disorder. Such treatment may indeed modify the symptoms, but that is not the same thing as curing the disorder.

One must additionally make a distinction between medical research and clinical medicine. They do not necessarily correlate. Medical research, whatever it chooses to study, plays by certain rules. Clinical medicine, on the other hand, tends to be less objective and often follows diagnostic and therapeutic trends despite the lack of evidence to support their validity.

Though physicians should lead the way to enlightenment by the exercise of good judgment and objectivity, they are frequently victims of the same prejudices held by laymen about things psychological and are equally uneducated about them. The degree of their psychological naiveté, including their inadequate knowledge of their own psyches, is astonishing, and more than a little scary.

The consequences of this medical failure have been catastrophic. It has spawned the major epidemics described earlier, and fostered numerous minor epidemics that once barely existed, like the whiplash syndrome, knee pain, foot pain, and shoulder pain. New and expensive therapeutic practices and whole new industries have been developed to treat these disorders, making it unlikely that enlightened change can be expected in the near future.

Let me emphasize that I know many physicians who are caring and do a wonderful job with their patients, surgeons among them. They are stars in the medical firmament. But because of the present climate in medicine, most of them cannot and will not make a psychosomatic diagnosis. Mindbody medicine is a world apart and has very few practitioners.

PSYCHIATRY AND PSYCHOSOMATIC MEDICINE

As pointed out earlier, official psychiatry has not recognized psychosomatic medicine for years. Even the term has been banished from the *Diagnostic and Statistical Manual of Mental Disorders* (DSM) and been replaced by the term *somatoform*. It is informative how the DSM deals with this matter. The introductory paragraph defines somatoform as follows:

> The essential features of this group of disorders are physical symptoms suggesting physical disorder (hence somatoform) for which there are no demonstrable organic findings or known physiologic mechanisms, and for which there is positive evidence, or a strong presumption, that the symptoms are linked to psychological factors or conflicts. Unlike Factitious Disorder or Malingering, the symptom production in Somatoform Disorders is not intentional, i.e., the person does not experience the sense of controlling the production of symptoms. Although the symptoms of Somatoform Disorders are "physical," the specific pathophysiologic processes involved are not demonstrable or understandable by existing laboratory procedures and are conceptualized by psychological constructs. For that reason they are classified as mental disorders.

What is particularly disturbing about this statement is that it may well apply to hysterical symptoms, but it certainly does not apply to psychosomatic disorders. Two phrases in the definition are of special importance: "physical symptoms suggesting physical disorder (hence somatoform) *for which there are no demonstrable organic findings or known physiologic mechanisms*" and "the specific pathophysiologic processes involved *are not demonstrable or understandable by existing laboratory procedures*" (italics mine).

These two phrases bring us to the heart of the matter for they represent *a matter of opinion on the part of the psychiatric community*, not a scientific construct. Put bluntly, the opinions of general psychiatry on the existence or nonexistence of psychosomatic disorders are irrelevant. Psychiatrists lack expertise in the domain of physical disorders, and therefore have no basis for an opinion as to whether a given set of symptoms represents a structurally induced or a psychosomatic condition. People with physical symptoms such as back pain or gastroesophageal reflux do not consult psychiatrists. One fails to see, then, how the writers of the DSM can have taken upon themselves the prerogative of deciding that psychosomatic disorders do not exist, as they have done in recent editions of that widely consulted reference work. It makes as little sense as it would for dermatologists to arbitrarily decide to render opinions on neurological disorders.

Regarding the manual's phrase "demonstrable organic findings," as long ago as 1888 Freud, working with patients with muscular rheumatism (known today as TMS), demonstrated the presence of pain on palpation (medical examination by touch), which is surely an organic finding. The disorder is clearly a mindbody process, with many demonstrable physical signs. The writers of the manual have been *unaware of*, or have *simply chosen to ignore* the evidence for the existence of psychosomatic disorders like TMS and the common gastrointestinal and allergic disorders described in this book.

Historically, there have been very few people qualified to judge whether a disorder is psychosomatic, and some of the best of them have been unaware of the most common of these conditions, the pain syndromes. As will be seen in chapter 2, Sigmund Freud described TMS but concluded it was "organic." Alfred Adler did not go into detail but stated that many physical symptoms were induced by the brain. Perhaps the best paper on the subject was "Psychogenic Regional Pain Alias Hysterical Pain," by Dr. Allan Walters, a highly respected Canadian neuropsychiatrist, published in the journal *Brain* in 1961. Walters described patients with pain that was clearly emotional in origin but who were not *hysterics*, as that term was used at the time. It is apparent that he was describing what we now call TMS.

It would appear that modern psychiatry has regressed back to the nineteenth century, when the predominant view of mental disorders was that they were either hereditary or due to brain disease. Freud had not yet introduced the idea that psychology, not physiology, was the important factor in mental disorders. So pervasive was the conventional view, however, that even Freud had trouble disavowing it. Now, despite evidence to the contrary, modern psychiatry suggests that the psyche does not induce emotional states like anxiety and depression and prefers to view them as chemically caused—back to the old nineteenth-century physiology again, albeit in a more sophisticated form. One cannot help but suspect that much of this is simply a repudiation of Freud, which can be dangerous and short sighted. It's true enough that Freud may have been in error about some details, but his basic ideas on the workings and importance of the unconscious are sound. Our experience with TMS makes that crystal clear.

In 1895, Josef Breuer and Sigmund Freud published *Studies on Hysteria*. It is a fitting bridge between this chapter and the next to mention two of the cases Freud described, for they recapitulate some issues just discussed, including Freud's description of what we now call TMS, his failure to recognize it as psychosomatic, the occurrence of a variety of psychogenic symptoms in one of the cases, and his and Breuer's pioneering concepts on the unconscious. The cases will then be examined in greater detail in chapter 2.

EMMY VON N AND ELISABETH VON R

Frau Emmy von N was a woman in her early forties who illustrates a concept suggested earlier: that the same psychology may give rise to a variety of psychogenic symptoms. First of all, she had emotional symptoms including anxiety, phobias, compulsive behaviors, delusions, and hallucinations. But she also had physical symptoms. Some were clearly hysterical in type; others were typical of what we see in patients with TMS, what was then called muscular rheumatism, a psychosomatic manifestation. So she had three of the psychogenic categories described earlier.

Fraulein Elisabeth von R was twenty-four when Freud first saw her. She had symptoms that were almost exclusively of the muscular rheumatism (TMS) type and a history typical of the cases I work with today. Here is what Freud said about muscular rheumatism:

> It seems probable that in the first instance these pains were rheumatic; that is to say, to give a definite sense to that much misused term, they were a kind which resides principally in the muscles, involves a marked sensitiveness to pressure and modification of consistency in the muscles, is at its most severe after a considerable period of rest and immobilization of the extremity (i.e. in the morning) is improved by practicing the painful movement and can be dissipated by massage. These myogenic pains, which are universally common, acquire great importance in neuropaths. They themselves regard them as nervous and are encouraged by their physicians, who are not in the habit of examining muscles by digital pressure. Such pains provide the material of countless neuralgias and so-called sciaticas, etc.

This is a brief but remarkable description of one of the many pain patterns of people with TMS, of whom my colleagues and I have seen literally thousands. Freud was a peerless observer. It is interesting that though he attributed all of the symptoms to muscle, he mentions neuralgias and sciaticas in his descriptions, both of which are nerve manifestations of TMS. Those who have read my books describing TMS will recognize his description. The pity is that he did not recognize that *the psyche initiated the process*. In this case he would have done well to heed the patients' family doctors, who said the symptoms were "nervous." He was fooled by the truly physical nature of the symptoms, the pain on digital pressure, which is one of the hallmarks of TMS.

Freud's view at the time was that the process was "organic"—that is, originating in the body, not the mind—because of the physical findings on examination. His view was entirely justified by the neuroscience of the time. He further believed that the psyche simply used the symptoms for a neurotic purpose. I think he would have discovered the truth had he continued to study physical manifestations, but he turned his attention to the neuroses and had very little to say about physical symptoms as his career developed.

The principle of emotional and physical equivalency plays out differently today than it did in Freud's time. Cases like Frau Emmy von N are unusual now, because hysterical signs or symptoms are rare, though I encountered one recently. The patient was a young woman in her twenties who described feeling as though her leg was sinking into the ground when she walked. One of the characteristics of a hysterical symptom is its bizarre, unreal quality, demonstrated classically in this young woman. Generally, people now tend to have either a physical or an affective symptom—either TMS (or one of its equivalents) or emotional manifestations like anxiety, depression, phobias, or obsessions. The medical profession recognizes the latter but not the former as psychological. The prevailing pattern depends on what is in vogue. Hysterical signs and symptoms are out of fashion; TMS is in, with all its variations like low back pain, "sciatica," neck and shoulder pain, "fibromyalgia," "carpal tunnel syndrome," knee pain, hip pain, and on and on. Gastrointestinal symptoms are also in vogue. Less commonly, a patient may have emotional and physical symptoms concurrently. I had such a patient recently. The young man came with a history of quite severe back pain of two years' duration. He did well in the program and became pain free in about three weeks. Shortly thereafter, he began to feel anxious and began to have some of his old stomach trouble again. This was the symptom imperative at work. The occurrence of two simultaneous psychogenic manifestations clearly suggested the need for psychotherapy. Either the severity of a symptom, emotional or physical, or the existence of two or more at the same time is an indication of the power of the unconscious conflict within. To extend this further, more serious physical disorders like the autoimmune, cardiovascular, or neoplastic disorders suggest more deeply repressed phenomena.

In our view, all of Frau Emmy's and Fraulein Elisabeth's symptoms, affective or physical, hysterical or psychosomatic, served the same purpose, that is, as a defense against powerful emotions in the unconscious that were striving to come to consciousness, or were being repressed because of their emotionally painful nature.

TWO

A BRIEF HISTORY OF PSYCHOSOMATIC MEDICINE

Psychosomatic medicine is a ghost, a set of ideas without a body. No one has ever really practiced psychosomatic medicine because its definition and scope have never been clearly established. It is my hope that this book will help remedy that situation and in the process make clear *that almost all of the common pain disorders that have afflicted millions through the years are psychosomatic.* That assertion will, of course, be vehemently disputed by almost everyone in medicine and psychiatry today, but I am confident that even a brief look at the history of the subject will settle the question.

The principal figures in the history of psychosomatosis are Jean-Martin Charcot, Josef Breuer, Sigmund Freud, Alfred Adler, Franz Alexander, and Allan Walters. All contributed in some fashion to the basic subject. They all made observations that help to explain psychosomatic phenomena, the most important of which was Freud's gift of knowledge of the unconscious, without which it would not be possible to understand emotionally induced physical symptoms.

CHARCOT

The story of psychosomatosis begins with Jean-Martin Charcot, the legendary French physician of the last half of the nineteenth century, who carried on his work at the Salpetriere Infirmary in Paris, starting as an intern circa 1856, then studying and lecturing for years as a volunteer. He occupied the Chair of Pathological Anatomy, then the Chair of Neuropathology from 1881 until his death in 1893.

Charcot's primary interest during the early years of his career was the systematic identification of neurologic disorders, but by the time Freud arrived to study under him in October 1885, he was devoting himself exclusively to research into the neuroses, particularly hysteria. In his characteristically meticulous fashion, Charcot had set about carefully describing the various manifestations of hysteria, and through his efforts had managed to elevate the clinical condition from a subject of ridicule to an established diagnosis in the classification of neuropathological disorders. It is entirely possible that the broad public attention to hysterical symptoms engendered by Charcot's work may have contributed to the high incidence of the disorder in his time. As we have already seen, psychogenic disorders have a tendency to spread in epidemic fashion when they are in vogue.

Charcot was a brilliant teacher, and he had a profound effect on Freud's career. As a result of the few months he studied with Charcot, Freud made the fateful decision to shift his focus of study from neuropathology to psychopathology, that is, from the study of the nervous system to the study of the mind. This crucial change of direction would in time lead directly to his theories on unconscious emotional processes that are fundamental to our understanding of psychosomatosis.

Though Charcot is revered by neurologists for his early work, I suspect that history will remember him best for his influence on Freud.

BREUER AND FREUD

Freud's friend Josef Breuer was the father of psychoanalysis. His famous clinical experience with "Anna O" from 1880 to 1882 has been described by Peter Gay (1988), in his biography of Freud, as "the founding case of psychoanalysis." When Freud returned to Vienna in 1886, having made the fateful decision to shift from neurology to psychology, he began to work closely with Breuer and to treat patients with hysteria. One of the results of this collaboration was the publication in 1895 of their book, *Studies on Hysteria*. Their friendship continued for years, with Freud the benefactor of professional, emotional, and even financial support from his older colleague. But the friendship later foundered on the rocks of conceptual differences.

Our interest here concerns what Breuer and Freud discovered about the unconscious and psychogenic physical symptoms in their work together. Their writings during this period are important because they establish a clear link between psychogenic disorders, then and now. The symptoms the two men were dealing with back in the nineteenth century were primarily hysterical, while today they are mostly psychosomatic, but the psychology responsible for both is identical. Further, it is interesting and intriguing that Freud briefly described what I have identified as the tension myositis syndrome (TMS). Freud was an excellent observer, and it is unfortunate that he failed to recognize that much of what he was observing was not of an hysterical nature, but psychosomatic.

In Freud's day, the treatment of hysterical symptoms included what he termed *aesthesiogenic methods*, such as electricity, the application of metals (such as copper bracelets), and the employment of skin irritants or magnets. For modern readers who might be tempted to adopt a condescending attitude toward such late-nineteenth-century medical technology, it should be noted that present day medicine has its own electrical treatments, including transcutaneous electric nerve stimulation (TENS). And copper bracelets and magnets are the rage with many back pain sufferers. *Plus ca change, plus c'est la meme chose.*

Freud noted that patients treated with such methods might appear to be "cured," but then soon after, they would simply develop new symptoms. He commented on the symptom substitution phenomenon—what we now call the symptom imperative—but did not understand its psychological significance.

Because of the importance of the symptom imperative, I'm going to repeat the rationale behind it. If the psyche has induced a physical symptom (such as back pain) or an emotional symptom (such as depression), which is then temporarily relieved in some fashion without dealing with the underlying emotional dynamic, the psyche will simply create another symptom to take its place. For example, if surgery is employed to relieve back pain due to TMS, it will prove to be only a placebo "cure," and similarly, if Prozac is used to treat depression, it will prove to be only a chemical "cure." In both cases, the patient will soon develop new symptoms. The TMS and the depression are not disorders in themselves; they are symptoms of unconscious conflicts and must be treated with psychotherapy to avoid the inevitable return of new symptoms. Occasionally, even those patients of mine who experience relief of symptoms from my cognitive-behavioral program (see chapter 4) will demonstrate the symptom imperative, in which case I recommend they begin psychotherapy.

Freud noted that a physical trauma frequently brought on hysterical symptoms either due to fright or "by the part of the body affected by the trauma becoming the seat of local hysteria." The same phenomenon is commonly observed in patients suffering from TMS. Often, the "trauma" is both remote in time and of a minor nature, like a fall on the ice and the subsequent onset of buttock pain months later. More commonly, patients with TMS will develop symptoms in association with what might be called *perceived trauma*, wherein pain comes on suddenly while the person is engaged in some benign physical activity like lifting a suitcase or swinging a tennis racquet. The activity is thought to be the cause of the pain, while in fact it is merely a trigger for the onset of the psychosomatic TMS. Much of what Freud labeled *hysterical* was no doubt psychosomatic.

More subtle contemporary examples of perceived trauma are the whiplash syndrome and repetitive motion "injuries," both described in chapter 1.

Though lacking scientific validation, a variety of symptoms have recently been attributed to "chronic Lyme disease," a disorder that may not even exist. This is still another example of a failure to recognize the psychosomatic nature of the symptoms.

Freud noted another example of the clinical similarity of hysteria and modern psychosomatic processes—that the hysteria was often combined with neurasthenia, or what is known today as chronic fatigue. The same combination occurs commonly and often simultaneously in the patients at our clinic.

The similarity between nineteenth-century patients with hysterical symptoms and the contemporary population of people with

TMS is striking. Charcot, Breuer, and Freud understood that such disorders originated in the psyche. Modern medicine, because it ignores their findings, is lost in a fruitless search for nonpsychogenic explanations, and so the epidemics continue.

THE CONSCIOUS AND UNCONSCIOUS MINDS

There is a section in *Studies on Hysteria* entitled "Unconscious Ideas and Ideas Inadmissible to Consciousness—Splitting of the Mind," written by Breuer. Today, we would substitute the word *emotions* for *ideas*, but that disagreement aside, the concept that we humans have two minds is very important to an understanding of TMS. It is clear that we are two different people—one of them conscious and the other unconscious.

How did Breuer and Freud see the split mind? In *Studies on Hysteria* the unconscious is described as a shadowy place inhabited by ideas that were not strong enough to be conscious, but which, if circumstances warranted, could be retrieved and brought into consciousness. (Freud would later assign this attribute to a part of the mind he labeled the *preconscious*, but it is apparent that in 1895 he had not yet developed that concept.)

The two men were puzzled, however, by the fact that an unconscious idea, though not sufficiently intense to become conscious, could be strong enough to induce motor paralysis. How could a weak idea produce such a strong effect? To answer this, they suggested that the pleasure or displeasure that the idea provoked—that is, the nature of its emotional content—might determine whether or not it could become conscious. As they saw it, an emotion could lead to a physical response, for example, a man who has a violent unconscious angry reaction to something during a meal and begins to vomit. Thereafter, eating called up the unconscious "memory" and he was once again nauseated. The analysis of this condition is described in a sentence of great interest: "This memory started the vomiting but did not appear clearly in consciousness, because it was now without affect, *whereas the vomiting absorbed the attention completely*" (italics mine).

The observation that a physical symptom can absorb the person's attention is very important. It is one of the fundamental concepts of the psychosomatic process, and my colleagues and I have observed it in thousands of people. Physical symptoms, of either the hysterical conversion or psychosomatic variety, are intended to divert attention from emotions in the unconscious so that they will not become overt and thereby known to the conscious mind.

Fundamental to Freud's understanding is that psychic phenomena result in *excitation*, a term found throughout his writing, implying a kind of energy that produces symptoms of one kind or another and that can be transferred from one sphere of activity to another for psychic reasons. Modern neuroscientists have described systems indicating extensive interaction between brain and body explaining how they can influence each other. But that is not the reason for psychogenic symptoms. Psychogenic symptoms are meant to be *protective distractions.*

Breuer and Freud refer to unconscious ideas, and to the emotions attached to those ideas, but it is the ideas that they consider to be of primary importance, not the emotions. They state that some of these ideas are inadmissible to consciousness and are therefore pathological.

And then Breuer makes the very important statement: "I do assert that the splitting of psychical activity . . . is present to a rudimentary degree in every major hysteria and that the liability and tendency to such dissociation is the basic phenomenon of the neuroses." In other words, the mind-split between conscious and unconscious is neurotic.

This is a very important concept, but it is one with which my colleagues and I strongly disagree. I have concluded that since *everyone* experiences psychosomatic symptoms, the mind-split is a universal human trait, not something associated only with the neuroses. Or to put it another way, we are all "neurotic." Therefore, neurotic is normal.

One thing is clear from a reading of *Studies on Hysteria*, and that is that Charcot, Breuer, and Freud were the first to recognize that pain could be psychogenic, that it could originate in the mind. To this day, that idea is still rejected by mainstream medicine, to the detriment of millions. Between then and now, only Alfred Adler, Franz Alexander, and Allan Walters, who based their work on clinical observation, have come to a similar conclusion.

Breuer wrote, "Since hallucinations of pain arise so easily in hysteria, we must posit an abnormal excitability of the apparatus concerned with the sensations of pain." This is the same idea of hysterical pain suggested in chapter 1, namely, that the pain is due to an exclusively intracerebral process.

At the end of that chapter, I looked briefly at two of Freud's cases from *Studies on Hysteria*. In his discussion of both Frau Emmy von N and Fraulein Elisabeth von R he described symptoms that we would classify as characteristic of TMS. Regarding Frau Emmy von N, Freud said, "I will venture in the first place to include pains among somatic symptoms. (Good!) So far as I can see, one set of Frau von N's pains were certainly determined organically by the slight modifications (of a rheumatic kind) in the muscles, tendons or fascia which cause so much more pain to neurotics than to normal people." TMS involves pain in the muscles, nerves, and tendons of normal people. It is not "organic" but classically psychosomatic.

About Fraulein Elisabeth von R, he asked, "Why was it that the patient's mental pain came to be represented by pains in the leg rather than elsewhere? The circumstances indicate that this somatic pain was not *created* [italics Freud's] by the neurosis but merely used, increased, and maintained by it. I may add at once that I have found a similar state of things in almost all the instances of hysterical pains into which I have been able to gain an insight. There has always been a genuine *organically-founded pain* [italics mine] present at the start. It is the commonest and most widespread human pains that seem to be the most often chosen to play a part in hysteria."

Here is one of the most important of Freud's errors as it relates to the world of psychosomatic medicine. He believed the pains associated with hysteria were "organic," and that the brain, which had played no part in producing the pains, simply used them for its own neurotic purposes. He did not recognize that such pains were actually created by the brain to serve a psychologically protective, benevolent purpose. There is a world of difference between these two concepts.

There will be further discussion of Fraulein Elisabeth's case later and a reinterpretation of Freud's conclusions.

Their misperceptions aside, Breuer and Freud made fundamental, far-reaching contributions to an understanding of psychogenic phenomena:

- They were aware of the unconscious and explored its nature, thus establishing the idea of the split mind and of the conflict that exists between the more intelligent, ethical, moral conscious mind and the childish, primitive unconscious mind.
- They recognized that hysterical symptoms were generated entirely in the brain without physiological alterations in the body, though they were experienced in the body.
- They first described what I call the *symptom imperative*, the tendency for symptoms to shift to other locations when

they have been relieved legitimately or factitiously, as with a placebo.

- They were aware of the excessive nature of psychogenic physical symptoms.
- They observed the clinical phenomenon whereby the psyche will often choose the site of previous physical injury as the locus for symptoms.
- They reported the fact that a patient might have more than one set of psychogenic symptoms simultaneously (e.g., hysterical pain and neurasthenia).
- They made the very important observation that psychogenic symptoms achieve their purpose by powerfully engaging the person's attention.
- Of even greater significance in understanding psychogenic physical processes, they realized that the psyche represses undesirable emotions. Unfortunately, they were not aware that the purpose of repression was to prevent problematic emotions from coming to consciousness.
- Along with Charcot, they were the first to recognize that pain could be psychogenic. They missed by a hair that it could be *psychosomatic* psychogenic.

Virtually all of contemporary medicine, including most of psychiatry, denies that the brain has the capacity to initiate physical, chemical, or neural changes in the body. But Adler, Walters, Alexander, and some of their contemporaries in the first half of the twentieth century were fully convinced that the brain had this power. That concept is essential to understanding psychosomatic disorders.

FREUD

Freud's monumental accomplishments have undergone a good deal of revisionist nitpicking in recent years, but I heartily endorse Jared Diamond's judgment in the February 2001 issue of *Natural History:* "Only two scientists within the last two centuries clearly qualify as irreplaceable: Charles Darwin and Sigmund Freud." To support his claim, Diamond goes on, "To begin with, Darwin and Freud were both multifaceted geniuses with many talents in common. Both were great observers, attuned to perceiving in familiar phenomena a significance that had escaped almost everyone else. Searching with insatiable curiosity for underlying explanations, both did far more than discover new facts or solve circumscribed problems, such as the structure of DNA; they synthesized knowledge from a wide range of fields and created new conceptual frameworks, large parts of which are still accepted today."

It is appropriate to designate Freud as the grandfather of psychosomatic medicine since his genius introduced us to the world of the unconscious mind, a contribution to medical science of inestimable importance. Psychosomatic processes begin in the unconscious and, though it has yet to be widely appreciated by either physical or psychiatric medicine, unconscious emotions are a potent factor in virtually all physical ills.

Unfortunately, Freud is not the father of psychosomatic medicine, for he did not realize that his patients' physical symptoms were induced by the brain to serve a psychological purpose, which is our definition of a psychosomatic disorder. He thought that physical symptoms were "organic," that is, the result of some physical disorder, and were merely being "used" by the psyche for a psychological purpose.

It was Freud who first pointed out that we are conscious of only a portion of our mental activities, and that much of our thinking and feeling takes place totally outside of our awareness, in what he called the unconscious.

Freud also defined three components of the human mind: the id, the ego, and the superego. The id is the childish, self-centered, primitive part of the mind. One might describe the ego as the captain, the chief executive, the decision maker of the mind. Freud saw the superego as the moral, responsible, ethical element of the psyche. The ego and superego operate in both the conscious and unconscious areas of the mind. In other words, we are aware of much—but not all—of the mental activities associated with these components. The id, which is the childish, selfish, primitive part of us, operates entirely in the unconscious, and we have no direct awareness of it.

It is appropriate to point out, as Freud does, that one cannot divide the mind into the neat compartments suggested by id, ego, and superego. They represent characteristics or tendencies of the mind, which, however, acts as a single unit, the action representing the sum of all the factors at work in the mind at any given moment. We should refer to this unit as the self, the individual. If we do, the id can be seen as the evolutionary core of the individual, while the ego and the superego represent later developments designed to enhance the chances of survival. The nature of the id—its power, its character, its emotional makeup—is central to any understanding of how mindbody medicine works.

EVOLUTION AND THE ID

Because it operates exclusively in the unconscious, and therefore outside our normal awareness, the id is not easy to observe in action. A rare and particularly poignant and dramatic glimpse of the id was described in an article by Dr. Paul Broks in the British periodical *Prospect*. Broks trained at a rehabilitation hospital for people with neurological disorders, and he recalled one of his patients there, a seventeen-year-old boy who had suffered massive brain damage as a result of a tragic accident. The boy's skull had been crushed and much of his cortex had been destroyed. Here is how he described the boy:

> Beneath this the face worked relentlessly, writhing with anger and dread. He would growl and grunt. Sometimes he would make a howling, wailing noise but, apart from occasional volleys of obscenity, he was incapable of speech.

This is a partial description of someone whose intelligent, cognitive, civilized mind, housed in the cortical mantle of the brain, has been destroyed and who is left only with primitive remnants deep in the substance of the brain.

> He sat contorted in his wheelchair, head turned sideways and back at an uncomfortable angle, limbs buckled with spasticity, a stream of saliva dribbling from the corner of his mouth. And due to a quirk of damage to his nervous system he was continuously troubled by a painful erection.

This is the unconscious mind, unmodified by the rational, civilized influence of the neocortex, that part of the human brain that has been added in the process of evolution. The neocortex, sometimes referred to as the cortical mantle, reflects the evolution of our species from its primitive beginnings. The older parts of the brain, responsible for this unfortunate boy's behavior, are deep in the substance of the brain, just above the brain stem. The only language available to this decorticated brain is automatic and primitive. (I recall a priest who had suffered damage to the cortical language areas of his brain from a stroke and when he attempted speech could only produce an embarrassing obscenity.)

Dr. Broks described a touching scene when the unfortunate boy was visited by his mother:

> I watched as she cradled his broken head in her arms. For the time that she was with him, but not much longer, an extraordinary transformation came over his face. It became still. The rage and the random mechanical twitching subsided, and he seemed to regain his humanity.

The boy's reaction to his mother's loving embrace was evidence that the primitive human brain, which was all he had left, is also inhabited by feelings of love and kinship. We should not be surprised at this, since even lower animals exhibit similar behaviors. But though these loving emotions are present in the human unconscious, they are not dominant. If they were, we would be in nirvana. The residual child, self-centered, narcissistic, dependent, seems to be more influential than the gentler emotions and, in many, the capacity for brutality predominates—which is why the world is the way it is.

What is quite remarkable is that most modern humans are largely unaware of this other self that exists in each of us and of the impact it has on every aspect of our lives. It is not generally realized that intelligence is not everything, that an intellectual genius may be an emotional baby or a monster. There is no correlation between intelligence and emotional maturity or balance. A case in point is the terrorist activity that dominates so much of the news media these days. Terrorists must be very intelligent to achieve what they do, but they are governed by powerful emotional drives that are neither rational nor humanitarian.

At a personal level there is a battle raging in the unconscious of every one of us between the residual child-primitive that Freud called the id, and the representatives of reason and morality he called the ego and the superego. This conflict is responsible for psychosomatic symptoms. As Freud said in one of his lectures, "To adopt a popular mode of speaking, we might say that the ego stands for reason and good sense while the id stands for untamed passions."

Just so, but humans are not yet fully tamed! We need only remember Hitler, Stalin, Pol Pot, the horrors of Rwanda, and September 11, 2001, to realize that reason is not yet in full control, if it ever will be.

The conflict we are seeing in contemporary humans is a brain-mind in transition, not fully governed by intelligence, and still under the influence of primitive, childish impulses, desires, and dicta. These negative attributes are what might be called the dregs of evolution, and their social-political and medical reality is all around us.

This mental dichotomy is responsible not only for the common pain disorder I have described in my books, but also a host of other medical disorders initiated by similar psychic processes (gastroesophageal reflux, irritable bowel syndrome, etc.). It is the basis for public health problems of enormous magnitude, but these matters appear to be completely ignored by contemporary medicine.

Freud based his concepts of the id on a study of his patients' dreams and an exploration of their neuroses. He saw it as having a dark, chaotic character, illogical, irrational, narcissistic, dependent, childish, primitive, and capable of simultaneously contradictory impulses. Of particular importance is the fact that it is timeless. "There is nothing in the id that corresponds to the idea of time," he wrote, "there is no recognition of the passage of time, and—a thing that is most remarkable and awaits consideration in philosophical thought—no alteration in its mental processes is produced by the passage of time. Wishful impulses which have never passed beyond the id, but impressions, too, which have been sunk into the id by repression, are virtually immortal; after the passage of decades they behave as though they had just occurred. They can only be recognized as belonging to the past. . . ."

This time factor is of great importance in understanding the psychology of mindbody disorders. In a previous book I described how Helen, one of my TMS patients, experienced a powerful psychosomatic reaction upon recalling a case of sexual abuse that had been expunged from her conscious memory for *over thirty years*. Not only are repressed *impressions* and wishful *impulses* preserved without change in the id, but so are any *emotions* generated in the unconscious. Hence, unconscious anger generated in the mind of a boy of ten will be alive and equally intense when he is forty as on the day it occurred.

Freud did not write that emotions could be generated in the unconscious. One gets the impression that his understanding was that thoughts, impulses, and feelings were generated in the conscious mind, and could then somehow be pushed into the unconscious and kept captive there by repression. One of Freud's biographers, Peter Gay, describes the unconscious as a maximum security prison where all the desperate criminals (i.e., dangerous feelings) are kept under heavy lock and key. The analogy is helpful in understanding the psychology of psychosomatic disorders, for it is the drive of these feelings to come to consciousness (the attempt of criminals to escape) that necessitates symptoms to prevent that from happening.

Freud observes this very thing: "We must rather attribute [to] the repressed [ideas] a strong upward drive, an impulsion to break through into consciousness." The unconscious mind, it would seem, wants to join the conscious mind.

Jonathan Lear, a philosopher-psychoanalyst, expressed a similar thought when he wrote of the mind's tendency to effect "a unification of thought and feeling."

I was forced to the same conclusion because I could find no other way to explain the development of the symptoms in TMS except to see them as a deliberate distraction designed to prevent the dangerous, painful emotions of the unconscious from becoming part of conscious experience. But there is a difference between this concept and Freud's ideas on the role of repression. He wrote at great length on the subject, equating repression with resistance. According to Freud, the repressed ideas were reprehensible, unacceptable, often loaded with sexual content, and therefore the purpose of repression/resistance was to keep them hidden and to thwart any attempt at analysis. Our studies of TMS lead us to conclude that repression serves a *protective purpose* since the repressed emotions, should we become aware of them, would in some way be dangerous to normal existence or be too emotionally painful to deal with.

Freud says that these dangerous inhabitants of the unconscious can be made conscious through the process of analysis. However, in our experience, many repressed feelings and impulses simply cannot be brought to consciousness. It is as though rage, narcissism, sadness, and feelings of dependency or inferiority are permanent residents of the unconscious.

NARCISSISM, NARCISSISTIC RAGE, AND INFERIORITY

It is essential to recognize the violent, brooding nature of the unconscious, and it is equally important to understand how it got that way. Freud made some cogent observations on this latter point in *Beyond the Pleasure Principle*, although he was not aware of the full implications of what he wrote:

> The early blossoming of infantile sexual life is doomed to extinction because its wishes are incompatible with reality and with the inadequate state of development which the child has reached. That blossoming comes to an end in the most distressing circumstances and to the accompaniment of the most painful feelings. Loss of love and failure leave behind them a permanent injury to self-regard in the form of a narcissistic scar, which in my opinion contributes more than anything else to the "sense of inferiority" which is so common in neurotics.

And later in the same section, in reference to the same infant-child:

> The lessening amount of affection he receives, the increasing demands of education, hard words and an occasional punishment—these show him at last the full extent to which he has been scorned. These are a few typical and recurring instances of the ways in which the love characteristic of the age of childhood is brought to a conclusion.

These observations help to explain the feelings of inferiority present in the unconscious mind, and by that we mean *everybody's* unconscious. My experience of working for many years with a very large cohort of patients with psychosomatic disorders supports the view that these feelings of inferiority are universal, and not restricted just to "neurotics."

It is likely that feelings of low self-regard are also enhanced when the developing infant-child compares itself to the giants all around it. Because of the timelessness of the unconscious, these feelings persist throughout life and are compensated in some people by the drives to be perfect and/or good and in others by aggressive behavior. This explanation for the existence of feelings of inadequacy in the unconscious is powerfully buttressed by clinical observation, as we will see when we come to examine the psychology of psychosomatic disorders. Feelings of inferiority play a crucial role in most people's symptoms.

EVOLUTION, THE EGO, AND THE SUPEREGO

Freud described the *id* as the fundamental self, the bedrock person inside each one of us, and the *ego* as a component that has come into being to protect it. "The ego is after all only a portion of the id, a portion that has been expediently modified by the proximity of the external world with its threat of danger." I would add that from a Darwinian perspective the ego evolved out of the id to save it from the extinction suffered by less fortunate evolutionary cousins. One might say the ego developed in response to the overarching evolutionary imperative: *to survive.*

Freud understood the need to observe and analyze the ego. He called it "the sense organ of the entire apparatus." The ego interprets the world for the id and protects it from that world. To fulfill its function, it must be rational, logical, and aware of time.

But, of course, the ego is also aware of the id's demands and reactions. In fact, Freud pictured the ego as being under siege from both the demands of the id and the pressure of everyday life. He believed these combined pressures led inevitably to anxiety in the ego.

And as if things were not complicated enough, the mind has developed still another trait—the *superego*—that Freud viewed in moral terms. As he saw it, the superego insists that you not only have to survive, but you have to survive as a successful, achieving individual. My own experience with psychosomatic disorders adds still another dimension to the superego's influence: you must not only be moral, you must be a saint. You must be *perfect* and *good.*

And how does the id react to such imperatives? The pressures we put on ourselves, and the workings of the superego, *infuriate* the id. All the narcissistic id wants is to gratify its desires for comfort, pleasure, and dependency, but instead it is being pressured to be a responsible adult. The result may be emotional pain, sadness, anger, and, cumulatively, rage. To the pain and anger generated in childhood, we now add the emotions arising from the conflict between the residual child-primitive in all of us and the pressures imposed by life—personal relationships, job, social obligations, and so on—and the superego. We shall describe the full flowering of this contest in chapter 3.

While the id is entirely unconscious, the ego and superego function in both the conscious and unconscious realms of mental-emotional life. Because it is the educator and guardian of the self, one must conclude that it is the ego, perhaps at the behest of the superego, that decides on the protective strategy of *repression*, which is then reinforced by psychosomatic symptoms. The ego is aware of dangerous goings-on in the unconscious and that these feelings are striving to come to consciousness, so it takes steps, sometimes quite dramatic steps to be sure, to see that the danger and emotional pain remain contained. My experience with TMS has convinced me that the purpose of this repression is to protect the individual, to prevent the painful, dangerous feelings from coming to consciousness and causing even greater distress. The psychosomatic symptoms that accompany this repression, while sometimes extremely distressing, are not some form of punishment but are generated to distract the conscious mind and therefore to assist the process of repression.

Put another way, painful or otherwise distressing psychosomatic symptoms are designed for self-preservation, not self-flagellation. This will become more apparent when their psychology is fully described. In the course of that description I shall reinterpret two of Freud's celebrated cases, that of Elisabeth von R, which we have previously touched on, and the controversial case of Dora. Freud believed that repression was the resistance of the patient to bring painful or embarrassing thoughts to consciousness during psychoanalysis. In his view, the repressed ideas were reprehensible and were hidden from the conscious mind by the ego and superego because of their unsavory nature. After all, the superego is the keeper of morality. In my view, the repressed feelings are painful and dangerous rather than bad, and the drive to repress them is motivated by the need to protect the total individual. The psychosomatic symptoms—whether pain, discomfort, depression, or whatever—are activated solely to reinforce repression and protect the person from mental pain or discomfort.

The broader view, that the superego stimulates unconscious anger by pushing for perfectionism and goodism, is consistent with the idea of repression as protection from feelings that are dangerous. What Freud called resistance is then seen as a response to great fear of those repressed feelings and an unconscious unwillingness to experience them, not because they are morally reprehensible, but because they are dangerous and painful.

ANXIETY AND REPRESSION

In discussing the relationship between anxiety and repression, Freud makes the point, ". . . first, that anxiety makes repression and not, as we used to think the other way around, and [second] that the instinctual situation which is feared goes back ultimately to an external situation of danger."

In my experience, the state of anxiety, which is perceived by the individual as a psychological malaise, is a *reaction to what is being repressed*, created by the ego as a distraction, much as it creates depression and physical pain for the same purpose. Anxiety is an equivalent of pain and depression. It, too, acts to assist repression. What the patient fears is not an external but an internal situation of malaise and danger—painful feelings and rage. The patient is not conscious of these feelings. The anxiety is free floating, generalized to all aspects of the person's life. Pain and depression may alternate with anxiety, making it quite clear that they serve the same psychological purpose. This is another example of the symptom imperative, and I have had numerous patients who have exhibited precisely such symptoms. Pain, anxiety, and depression are not symptomatic of illness or disease. They are all part of the normal reaction to frightening unconscious phenomena.

AGGRESSION AND SELF-PUNISHMENT

Another aspect of Freud's studies dealing with the relationship of unconscious mental activity to physical symptoms involved his investigation of what he called the aggressive instinct and its twin, self-punishment. Freud considered self-punishment to be the aggressive instinct turned inward and described it as "a piece of aggressiveness that has been internalized and taken over by the superego." He cites the case of a woman who had suffered great disability for many years from what he called a "complex of symptoms," and from which he eventually liberated her. Since he does not specify the symptoms, we presume they were emotional rather than physical. However, when she attempted to become active after her recovery, she met with a variety of "accidents" that might have been psychologically induced, and she began to have "vegetative symptoms," including catarrhs, sore throats, influenza conditions, and rheumatic swellings. He believed these symptoms were the result of an "unconscious need for punishment."

My interpretation would be quite different. First, I see the shifting of symptoms as simply an example of the symptom imperative, which Freud himself had observed elsewhere. Since frightening unconscious rage continued to exist, so did the need for defensive symptoms. It is abundantly clear that whatever Freud did to alleviate her original symptoms, he did not get to the heart of the matter, which was the existence of those unconscious feelings. I shall repeat again and again, psychosomatic symptoms serve a *protective* rather than a *punishing* purpose. It would have helped if he had known that both the "accidents" and the vegetative symptoms he described were psychosomatic and were merely replacements for the symptoms he had alleviated. They were induced by the psyche for defensive purposes, not "an unconscious need for punishment."

In my view, aggression is not genetically built in, but rather the result of the unconscious rage that appears to be universal and accounts for the manifestations of aggression we see all around us. *What is genetically built in are the functional components of the mind—the id, the ego, and the superego, developed over eons by evolution—and the myriad manifestations of this mind, both good and bad.*

Humanity's debt to Freud is enormous. These remarks are not meant in any way to discredit him. They are reinterpretations of his brilliant observations, based on a broad clinical experience. I think he would have approved.

ADLER

Alfred Adler was a young practicing physician when Freud invited him in 1902 to join his psychoanalytic circle. He soon became an important member of the group, highly esteemed by Freud and designated by him as his successor to the presidency of the Vienna Psychoanalytic Society. However, conceptual differences developed and in 1911 Adler separated from Freud, organized his own society, later to take the name of Individual Psychology, and founded his own journal, the *Zeitschrift fur Individualpsychologie*.

One of the most important of Adler's ideas, with which I am in full agreement, is that people's unconscious feelings of inferiority are innate and universal, characteristic of both the "neurotic" and the normal. Adler spoke of *feelings* in contrast to Freud, who referred to unconscious *thoughts* and *ideas*. This is one of the differences that probably led to their split. Adler further postulated that feelings of inferiority stimulated a striving for superiority, for perfection and high accomplishment. His concepts of the reasons for striving toward such perfection went beyond the purely personal. He saw these strivings as being motivated in the normal individual by social interest, in the need to advance the common good. He even viewed this trend as a kind of societal evolution, with God representing the ultimate. "In God's nature religious mankind perceives the way to height," he wrote. According to Adler, the neurotic is motivated entirely by self-interest, having no "goal based on interest in reality, on interest in others, and on interest in cooperation."

Much of Adler's Individual Psychology is based on his view that humans cannot be properly understood unless they are viewed in a social context. "In order to understand what goes on in an individual," he wrote, "it is necessary to consider his attitude toward his fellow men." And at another point he observed, "Never has man appeared otherwise than in society."

Adler's psychology is a *social* psychology, while ours—psychosomatic psychology—is rooted in psychodynamics and neurophysiology: the superego versus the id, the neocortex versus the brain stem and hypothalamus. Human beings are, of course, both private and social individuals, and our behavior is influenced by everything that goes on in our lives. The environment is obviously important, but there is a greater role for the unconscious in every aspect of our lives than has heretofore been recognized, including the physical as well as mental and emotional lives. One cannot evaluate someone's goals, ambitions, achievements, personal relationships, social interactions, or physical or mental health without knowing what's going on in the unconscious.

From my point of view, it is likely that Adler's most significant contribution to our understanding of psychology was his recognition of the relationship of the psyche to physical symptoms. Freud, Adler's senior and teacher, asserted that physical symptoms such as pain, cough, and gastrointestinal disturbances were "organic"; that is, they were based on some disease process and were simply "used" by the psyche to serve a neurotic purpose. Adler's views on the subject indicate that *he was the first to recognize that the psyche could induce physical symptoms by initiating physiological pathology.*

He wrote, "The mind is able to activate the physical conditions. The emotions and their physical expressions tell us how the mind is acting and reacting in a situation it interprets as favorable or unfavorable." Bravo!

"A mental tension affects both the central nervous system and the autonomic nervous system," Adler observed. And more. "The body, through the autonomic nervous system, the vagus nerve, and endocrine variations, is set into movement which can manifest itself in alterations of the blood circulation, of the secretions, of the muscle tonus, and of almost all the organs."

There you have it: the pathophysiology of TMS and its equivalents described by Alfred Adler. He would have been delighted and astonished to learn of the elaborate peptide network that connects brain and body, validating his psychosomatic concepts.

What a wonder to contemplate: Freud described TMS in 1888, and Adler its physiology, circa 1911.

In his discussion of actual psychosomatic disorders, Adler identifies a variety of physical states and conditions, including:

- Immediate reactions, such as blushing, perspiring, and rapid heart rate (I consider these psychogenic, but not psychosomatic.)
- A group including headaches, bowel irregularities, and frequent urination (I categorize these as TMS and its equivalents, and definitely psychosomatic.)
- The feeling of a lump in the throat known as globus hystericus (This is clearly a hysterical disorder rather than psychosomatic.)
- Autoimmune disorders such as thyroid conditions
- Structural aberrations like lateral curvature of the spine (scoliosis) and flatfeet (I do not consider such structural abnormalities to be psychosomatic.)

Though his ideas on the meaning of psychosomatic symptoms differ from mine, Adler was the first to recognize that the brain could initiate physical symptoms that were not hysterical. He referred to "organ dialect," the body talking, so to speak. He also believed that symptoms tended to occur in "inferior organs" and that there was a symbolic reason for the particular organ or system chosen. Our experience does not support the idea of organ inferiority and Adler's reasons for particular organs being chosen for symptoms, but these matters are of secondary importance to his recognition of the basic nature of the psychosomatic process.

Adler saw emotions as accentuations of character traits. He thought conscious states like anger, sorrow, or fear related to a person's goals, lifestyles, and the like. He appeared not to put forth the concept of unconscious emotions. However, he provides a wonderful quote: "We shall generally find unadmitted rage or humiliation behind attacks of migraine or habitual headaches, and, with some people anger results in trigeminal neuralgia or fits of an epileptic nature."

According to Adler, the rage is conscious but unadmitted. Another quotation makes this clear: "The unconscious is nothing other than that which we have been unable to formulate in clear concepts. It is not a matter of concepts hiding away in some unconscious or subconscious recesses of our minds, but of parts of our consciousness, the significance of which we have not fully understood."

I strongly disagree. My understanding of the unconscious is precisely what Adler says it is not. The heart of the psychosomatic process is to keep painful and dangerous emotions repressed and hidden in the unconscious, because these are emotions that would wreak havoc were they allowed to become conscious. The unconscious is a domain, a realm that is home to a variety of concepts, thoughts, ideas, feelings, traits, and tendencies. Some are positive, pleasant, and socially acceptable (as opposed to antisocial) and some are negative (e.g., feelings of inferiority). Some are violent, some obscene, some childish (e.g., narcissism and dependency), some dangerous and threatening (e.g., rage), and some are simply too painful and sad to be consciously experienced.

Adler's recognition of the relationship of rage to migraine and other disorders is impressive. It is quite possible that his assertion that trigeminal neuralgia is psychosomatic is correct. I have speculated that not only trigeminal neuralgia but Bell's Palsy and other mononeuropathies, such as that of the long thoracic nerve of Bell, are the result of psychically induced local ischemia (mild oxygen deprivation).

To Adler there were good emotions, like joy or sympathy, which led to socially laudable goals, and bad emotions, like anger, fear, or sorrow, that are characteristic of the neurotic. As with Freud, he considered neurotics to be sick people. Neither Freud nor Adler had the concept that we are all potential neurotics, which is why psychosomatic symptoms are universal.

Freud and Adler agreed that repression and symptoms were both defenses, but for Freud their purpose was the "protection of the ego against instinctual demands," while Adler saw them as defenses for protecting the self-esteem from the external demands and pressures of life. Our experience makes room for both interpretations and sees repression and symptoms as reactions to troubling emotions that are stimulated by both Freud's instinctual demands and Adler's demands of everyday life.

I found it of great interest that Adler thought the neurotic state was generated by a need to avoid "a greater evil," namely, *to prevent one's worthlessness from being disclosed.* Our experience with thousands of TMS patients leads us to find that the mind considers that the greater evil would be *the conscious experience of emotional pain and rage.* Another interpretation is that conscious feelings of worthlessness stimulate the drive to be perfect and good that, in turn, stimulates unconscious rage, resulting in a neurotic state.

Another Adlerian observation parallels our experience with TMS patients: he found those patients with feelings of worthlessness to be extremely sensitive to criticism. This appears to be a reaction to intense feelings of inferiority.

As can be seen, there is considerable similarity between Adlerian psychology and my own. But here's a reinterpretation of one of Adler's cases that highlights the differences between the two: the patient was a twenty-five-year-old woman who suffered violent attacks of anxiety when her husband was late in coming home from work. Adler concluded that she had never received much attention at home prior to her marriage, that her husband tended to overindulge her, and that the attacks were designed to counter his interest in his business and redirect it to her.

My own analysis would be that the young woman was unconsciously in a rage because her husband was more interested in his business than he was in her, and further, that she had been deprived of his attentions when he was late.

I had a patient who would get violent migraine headaches when her husband was late in returning from a hunting or fishing expedition. In her case, the unconscious rage that brought on her headaches probably resulted from her fear that something had happened to him, and was further intensified by the idea that he had exposed her to such worry.

To summarize Adler's ideas that are pertinent to understanding psychosomatic disorders:

- Unconscious feelings of inferiority are universal.
- Feelings of inferiority stimulate the drive for superiority and perfection.
- The brain has the capacity to induce physical symptoms like pain, cough, or gastrointestinal disturbances, when motivated by psychic phenomena.
- Symptoms may be created by autonomic activity (e.g., in the circulatory system) or by the endocrine network. The basis for migraine headache, "habitual" headache, and trigeminal neuralgia is rage.

- Psychogenic symptoms are defenses designed to protect one's self-esteem from the pressures of life.
- Sensitivity to criticism is prominent in patients with low self-esteem.

Because I was trained as a physical physician, rather than as a psychiatrist, I was not aware of Adler's work relating to psychosomatosis until long after I had formulated my own concepts on the nature of the psychosomatic process. I am pleased that so many of his conclusions paralleled my own.

ALEXANDER

For half a century after its genesis in Vienna, the concept of psychosomatic medicine continued to develop and attract adherents. During those years, one of the foremost leaders in the movement was Franz Alexander, who had been one of Freud's students and was the founder of the Chicago Institute for Psychoanalysis. In his book, *Psychosomatic Medicine* (1950), Alexander hailed the developing awareness of the psychosomatic reality within the medical community and prophesied a great future for what he perceived to be an important new tool for the healing profession. Sadly, that great future never arrived, the enlightenment he described did not flower, and the vital new diagnostic tool he envisioned never developed. Ironically, one can date the medical community's gradual moving away from the ideas he espoused to around 1950, the year his book was published, and coincidentally, the year of my graduation from medical school. I can recall virtually no teaching on the subject.

The scope of Alexander's work is impressive. He studied the role of emotions in gastrointestinal, respiratory, cardiovascular, dermatologic, metabolic, and endocrine disorders and rheumatoid arthritis.

Alexander and a group of neurologists and psychoanalysts published extensively on these disorders during the first half of the twentieth century, and this no doubt was responsible for his optimism about the future of psychosomatic medicine.

Alexander believed that emotions played a role in all illness, even in those cases where they were not necessarily the cause of the illness. And he insisted that it was as important to identify with precision the emotional components of a disorder (e.g., cardiac) as the physiological ones.

Alexander anticipated the current preoccupation with chemical psychiatry, and warned, "A biochemical formula describing a receptive longing somewhere in the cortex will never account for the interpersonal circumstances under which this longing arose or became intensified." In other words, a symptom and its cause can be very different things. He would have been appalled at the contemporary tendency to explain almost all medical phenomena on physical, chemical, or genetic grounds. It is naive to think that linking a brain chemical to a behavioral state establishes the cause of that state. Alleviating depression with a powerful drug does not eliminate the reason for depression; it merely treats the symptom.

CHRONIC FATIGUE SYNDROME

Alexander's discussion of a patient suffering from what is today called chronic fatigue syndrome (CFS) is instructive. He described the case of a thirty-one-year-old writer who had been in long-term psychoanalytic treatment. He had suffered chronic fatigue and acute attacks of exhaustion since the age of seventeen:

> The patient was an unwanted child, small at birth, and remained physically underdeveloped his whole life. He suffered from inferi-

> ority feelings because of his small stature and weakness. The parents' marriage was unhappy; the father drank heavily and neglected and abused his family. All his life the patient remained very close to his sister, who was three years younger than he. He developed a tremendous fear of his father; it stood out in his memory that when his father caught him masturbating he threatened him with insanity. When he was only eight years old his father insisted on his working around the house, peddling articles, or caddying, all of which he did only under internal protest. When he was ten, he had a sexual relationship with his sister. To run away with her into "the never-never land" was for a long time a cherished regressive fantasy. In school he gradually withdrew from activities; he was afraid of both teachers and students. He changed colleges several times and at twenty-three started to work in a factory. For a while, he also worked as a sailor and as a day laborer, but finally he began to write and do editorial work, for which he had talent. He was successful and able to do his work.

Alexander's interpretation of this case led him to make a very important statement: "This psychodynamic constellation—the conflict between passive, dependent wishes and reactive aggressive ambition—is widespread, if not universal, in our civilization and can hardly be claimed as a specific explanation of this type of fatigue syndrome."

We would interpret the situation somewhat differently. We would say that this patient was in perpetual, ever-increasing unconscious hurt, suffering from a combination of emotional pain, sadness, and rage dating back to early childhood, and that these were responses to his terrible family situation, his fear of his father, his small size, and his strong feelings of inferiority, all of which were intensified by his father's treatment of him. And then, to further feed these feelings, he put pressure on himself to accomplish something, the desire that Alexander described as his "reactive aggressive ambition." His physical symptoms were designed to spare him from experiencing these feelings consciously. His fantasies reflected the continuing effort of the unconscious child in him (as in all of us) to escape the realities of his existence. His physical symptoms were unimportant, and therefore treating them medically was of limited value. Our work with patients with TMS consistently demonstrates that the pain problem is *psychological*, not *physical*.

As with all psychosomatic maladies, CFS is a mystery to the modern medical world. In a report published in 1996, a group representing three of Great Britain's royal colleges was unable to identify a specific cause for the disorder, but observed that 75 percent of the population with CFS suffered from one or more of the following: depression, sleep disturbance, poor concentration, agitation, feelings of worthlessness, guilt, suicidal thoughts, and appetite or weight changes, as well as anxiety or physical symptoms related to anxiety or depression. Accordingly, as treatment they recommended increased physical activity and cognitive-behavioral psychotherapy.

I have seen many patients with this disorder who came to my clinic primarily because of pain complaints and who did well under my care. I strongly believe CFS to be a psychosomatic disorder, and this belief is supported by my success in treating a very large number of cases. A further indication that we are right is that many people suffering from CFS have recovered from it simply by studying one of my books.

MIGRAINE AND HYPERTENSION

Alexander's observations of migraine and hypertension provide important insights into the nature of many psychosomatic disorders.

He and others contributing to the literature at the time found that "inhibited hostile tendencies" were important in the genesis of hypertension. Alexander found this entirely plausible since it had already been demonstrated that fear and rage elevated blood pressure in laboratory animals. In fact, the term *repressed rage* was used by many authors describing patients with hypertension and migraine. Alexander was particularly impressed by the rage factor in migraine, noting that he had observed the immediate cessation of headache as soon as his patient discharged his rage with a sharp verbal outburst of vulgarity. Others at the time reported that migraine patients tended to be perfectionistic, ambitious, competitive, rigid, and unable to delegate responsibility. Another noted the inhibited, "goody-goody" behavior of migraine sufferers. Our own studies support their findings that the twin characteristics of perfectionism and goodism are major factors in the genesis of TMS and its equivalents, and that those same characteristics are enraging to something in the unconscious.

It should be noted that hostility, aggression, and rage are not synonymous. Hostility and aggression are observable and are the *consequences* of unconscious rage. They are not built in. Another misconception is that rage is always conscious and suppressed. Rage may be overt, it may be conscious and suppressed, or it may be unconscious and repressed, completely outside of the person's awareness. In the case of unconscious rage, it is likely that it is not a response to hostile-aggressive needs but a reaction to pressures imposed upon the individual by the vicissitudes of life or, even more enraging, by the person himself. That idea was supported by Alexander, who noted that patients who developed hypertension typically gave a history of having been very aggressive during early life, then suddenly, often during puberty, finding that their aggressiveness made them unpopular, became meek and easily intimidated. They reported that they had to make a conscious effort to control themselves.

One could say that these patients' aggression was a reflection of unconscious rage, that social imperatives required that they consciously control their aggressive behavior, thereby increasing the internal rage to the point where it now needed a physical disorder to prevent its explosion into consciousness, hence the hypertension. In our clinic we see these people not as undesirable hostile-aggressives, but as *victims of circumstance*, victims of the pressures of life and the pressures they put on themselves. They are not aggressors; they are angry. They are not sinners; they are being sinned against. If they appear to be aggressive at this point, it is a reaction to feelings of impotence and/or the need to vent their unconscious rage. It is likely that if they express anger, it will be what is known by psychologists as "displaced anger"; it will not be the rage that is repressed.

RHEUMATOID ARTHRITIS

Norman Cousins was an important literary figure and editor. He also suffered from rheumatoid arthritis (RA). Ever since the publication of his book *Anatomy of an Illness* in 1979, a portion of the lay public, at least, has accepted the fact that rheumatoid arthritis is somehow related to emotions. The book described Mr. Cousins's failure to respond to conventional medical treatment of an acute attack and his eventual recovery by recognizing the role of emotions in the disorder, and immersing himself in humorous books and films as an antidote to "bad feelings." To my knowledge, most rheumatologists continue to treat RA with medications and make no attempt to address emotional issues.

To Alexander and his colleagues the psychological basis for RA was abundantly clear. In his book he refers to studies which found that the disorder occurred primarily in women who, as adolescents, had a predilection for outdoor activities and competitive sports and who, as adults, suppressed all emotional expression and had a compulsion to strongly control their environment, including the lives of their husbands and children. At the same time they were extremely caring and protective of their families and exhibited such a powerful need to care for others that Alexander characterized it as masochistic. Sexually, they seemed to reject the feminine role and tended to select passive, compliant men as husbands. Alexander makes this statement: "The general psychodynamic background in all cases is a chronic inhibited aggressive state, a rebellion against any form of outside or inside pressure, against being controlled by other persons or against the inhibitory influence of their own hypersensitive consciences."

This observation closely parallels my own findings, and with one change is an excellent description of many people suffering from TMS. For "chronic inhibited aggressive state" I would substitute "chronic unconscious rage." The sources of the rage are the outside and inside *pressures* (the perfect word), the threat of control by others and the pressures brought to bear by their own superegos. I could not have composed a more succinct statement.

Anyone studying psychosomatic disorders is bound to be impressed by the powerful influence of childhood experiences. In his work on RA, Alexander repeatedly finds a strong, domineering, demanding mother and a compliant father, leading to fear of the mother coupled with dependence on her and an unexpressed desire to rebel. Once more my experience would suggest a small change: fear of and dependence on the mother leading to *unconscious rage* at her.

If Alexander were alive today and aware of the rheumatoid factor associated with RA, I believe he would no doubt agree with the observation that deep-seated rage somehow plays an important role in the genesis of the disorder, although it is still to be determined how this emotion stimulates the production of the rheumatoid factor and why the psyche chooses this rather than some other physical manifestation.

Alexander, along with his colleagues and contemporaries, observed many of the psychic phenomena that have led to my theories of psychosomatic psychology, including childhood abuse, hostility, aggression, rage, feelings of inferiority, passive dependent wishes, and ambitious drive. All are mentioned by Alexander, and all are part of the psychosomatic process. He occupies an important place in the history of psychosomatic medicine. He was at a disadvantage in not having had access to the millions now afflicted by psychosomatic disorders, the common pain problems, or he might well have developed strong evidence of his contention that mindbody maladies are universal.

WALTERS

On July 14, 1959, Allan Walters, a distinguished Canadian neuropsychiatrist, delivered a presidential address to the Canadian Neurological Society entitled "Psychogenic Regional Pain Alias Hysterical Pain." The paper was published in the journal *Brain* in March 1961. (Nowadays, it is highly unlikely that a neuropsychiatrist would be president of a neurological society, or that a paper with that title would be accepted for publication in a neurological journal. Sad but true.)

In my view, Walters's ideas are important because he is the only one after Franz Alexander to make a significant contribution to our knowledge of psychosomatic phenomena. In his paper he concluded that the patients about whom he was writing had psychogenic pain, but since only hysterical pain fit that category and many of his patients were not classic hysterics, he proposed to describe their disorders as *psychogenic regional pain* (*PRP*), a term that would include the entire spectrum of people with pain of psychic origin. He speculated about, but could not identify, what was responsible for their symptoms, beyond the recognition that they were psychological in origin. The accuracy of his diagnosis was confirmed in many cases by the patients' prompt return to normal when the psychology behind the symptoms was identified.

Walters identified three types of psychogenic pain: psychogenic magnification of physical pain; psychogenic muscular pain (tension headache, "fibrositis," spastic colon); and PRP, which he defined as pain for which no local reason could be detected but which was clearly psychological in its evocation. The last group was the subject of his paper.

Walters described the cases of 430 patients. They ranged in age from twenty to sixty. The sex ratio was three males to seven females. About two thirds had been referred by what were then called general practitioners and medical specialists. The other third came from surgical specialists looking for help in diagnosis. How times have changed! The primary care physicians and internists of today would refer such patients to neurologists, orthopedists, neurosurgeons, or pain specialists. Some would be referred to chiropractors or acupuncturists. And there would be a broader age range. Today, we occasionally get teenagers, and older people in their sixties, seventies, and even eighties are not uncommon. The sex ratio would be different, as well. It is about fifty-fifty nowadays.

It is interesting to compare and contrast the regions of the body involved then and now. Of the 430 cases covered by Walters, 185 involved the head and neck, 133 the chest and upper limbs, and only 112 the lower back and lower limbs. The remainder were scattered over the rest of the body. Nowadays, the two major sites would be reversed, and the low back and legs would be the primary site. This is another instance in which the patterns of psychosomatic disorders depend on what is in vogue, and helps explain why disorders like "fibromyalgia" and "carpal tunnel syndrome" assumed such epidemic proportions in the short span of fifteen years at the end of the twentieth century.

Walters's description of the quality of the pain his patients suffered is of great interest. While some patients used the language of metaphor or simile ("My head feels squeezed in a vice," "My knee feels as if someone is twisting it unbearably"), others used simple terms like *burning*, *aching*, or *shooting*. It must be remembered that he was under the impression that these patients had pain patterns similar to those described by Charcot, Breuer, and Freud. He recognized that they were not typical hysterics, and coined the new term *PRP* to describe them. But he was not aware of the specifics of their disorder. What neither Walters nor his predecessors knew (and what the medical community still does not know) is that most of those patients were suffering from a disorder initiated in the brain known in Freud's time as "muscular rheumatism" and nowadays are included in a group of disorders that constitute mindbody syndromes, namely TMS. I have heard similar descriptions in countless numbers of patients in my own clinic. TMS is *psychosomatic psychogenic*, in which the symptoms include physical alteration in some tissue, as opposed to *hysterical psychogenic*, where there is no physiologic change in peripheral tissues, but where very real pain occurs due to the stimulation of appropriate brain nuclei. Clearly, Walters had both types of patients in the group studied but was unaware of the existence of TMS.

The sites of the pain made no sense to Walters, which is why he used the term *regional* to describe them. He did not know that the brain can locate the pain anywhere in the body through the stimulation of specific brain nuclei, in cases of hysterical psychogenic pain, or by reducing the blood flow to particular muscles, nerves, or tendons in the case of TMS. One of the cases he described was that of a thirty-seven-year-old man who had pain that extended from the middle of his anterior thorax on the left side down to include his abdominal wall as well as a small segment of his posterior thorax on the same side. It is likely that the pain was the result of mild ischemia of nerve roots exiting the spinal cord (in which case it was TMS), or the stimulation of thalamic sites corresponding to those nerves in the thalamus (in which case it was hysterical psychogenic pain).

Another case, that of a forty-year-old woman, was diagnosed as situational depression and conversion hysteria, though he was greatly disturbed by the fact that the patient didn't exhibit the typical "belle indifference" so characteristic of the hysterical patient, which is why he included her in the category he called PRP. Both the symptoms and the location of her pain, the entire left arm and hand and a portion of the anterior upper thorax, make her a clear case of TMS, most likely a disturbance of numerous spinal nerve roots (polyradiculopathy). We have seen thousands of sufferers with this configuration of pain over the years. The patient recovered when Walters reassured her that she was not suffering a dread disease and helped her work out a stress reduction program, which must have convinced her that her pain was psychogenic in origin.

His course of treatment was similar to my therapeutic program, which begins with teaching patients that their pain is the result of an essentially benign process initiated by the brain, and not due to the conventional idea of a bodily disorder, and then helping them to understand the psychology of how and why the brain does what it does.

As pointed out in the first chapter, one can distinguish hysterical from psychosomatic symptoms by noting the quality of the symptoms. If they are truly bizarre, either in the patient's description or the anatomical manifestation, they are hysterical. Such symptoms include hysterical blindness, loss of voice (aphonia), and paralysis or numbness (anesthesia) of an entire limb. By contrast, psychosomatic symptoms are logical and clearly relate to specific muscles, nerves, or tendons or particular organs or systems (e.g., stomach, colon, urinary bladder, skin).

Another point previously noted is worth reiterating here. The choice of symptom, either hysterical or psychosomatic, depends primarily on what is in vogue at the time. Both hysterical and psychosomatic disorders existed at the time of Charcot, Breuer, and Freud, but hysterical symptoms were apparently more common than the psychosomatic ones, although Freud states that muscular rheumatism, which is what we call TMS, was common. Walters's data suggest that by the time of his studies TMS was more common than hysteria. In effect, it makes no difference what the psychogenic manifestation is, as long as it is recognized as psychogenic. That is why Walters was successful in his treatment of these patients—he knew the most important fact, which was that their symptoms were brain induced.

In describing the symptoms of PRP, Walters noted five kinds of physical signs:

1. Motor deficits

2. Tenderness

3. Sensory deficits

4. Changes in contact reflexes, like the cornea (rare)

5. Vegetative changes over the body surface (rare)

The first three are almost universal in patients with TMS.

In Walters's discussion of the psychopathology of PRP he concludes that symptoms may be produced:

- As a direct emotional expression
- By a substitutive process of conversion alone
- By a substitutive process of conversion with conversion hysteria
- By unknown processes

How these findings compare with the psychopathology of TMS will be covered in detail in chapter 3.

Walters was clearly ahead of his time. I became aware of his paper years ago in the early stages of my work with TMS and was heartened by the knowledge that there was a kindred soul out there.

Since Walters, there has been some speculation among psychoanalysts about psychosomatic medicine. While some theories have been developed, virtually nothing has been accomplished. Psychosomatic medicine is not part of the mainstream of medical practice, and as a result, the theories of these analysts have been of little value.

I must repeat what I have said before: no theory of mindbody medicine can be considered valid if it does not include the vast population of people with the pain complaints that are currently of epidemic proportions in the United States and most of the Western world, the ones we identify as TMS.

A final note on the history that has just been reviewed: Most theories, including those of Freud and his followers, considered psychosomatic manifestations as a form of illness representing *defective personalities.* I strongly disagree. *Psychosomatic phenomena are not a form of illness. They must be seen as part of the human condition—to which everyone is susceptible.* They include a wide range of disorders, some very serious and even life threatening, but our view is that they may all be traced back to the primeval conflict between our two minds, the unconscious and the conscious, the id and the ego and superego, the ancient "paleomammalian mind" and the modern "neomammalian mind," each mind reacting in the only way it knows to the pressures of daily life.

With these ideas as background, we are ready to move beyond the history of mindbody medicine to an examination of how it operates in today's world.

THREE

THE PSYCHOLOGY OF PSYCHOSOMATIC DISORDERS

Wordsworth wrote, "the Child is father of the Man," and that poetic truth has been validated by scientific observation many times in the last century. Every working day, at our clinic in the Rusk Institute of Rehabilitation Medicine at the NYU Medical Center, we have the opportunity to see and study how childhood experiences and the child in all of us—the timeless, primitive, unconscious mind that operates totally beyond our conscious awareness—continues to influence, and even shape, our adult selves.

All of my patients' case histories are interesting, but some are more interesting than others. I particularly remember the twenty-eight-year-old engineer who arrived at my office one morning and told me the story of his low back and leg pain. He had been suffering with it for over eight months. His condition was extremely painful, and he had gone the usual route of seeing a variety of doctors, trying a number of treatments, all without improvement. His MRI showed a herniated disk, which his doctors presumed to be the cause of his pain. Inevitably, the persistence of the pain after the failure of conservative treatment brought the usual recommendation for surgery. He might have gone that route, but it was around that time that he read my book, *Healing Back Pain* (1991), and decided that the disorder described in the book, TMS, might be the cause of his pain, so he made an appointment to see me.

One of the things he told me during the consultation was that he had an extremely responsible job, one in which he supervised four people, a couple of them older than himself, and that he found the job very burdensome. He then said that about two weeks before our scheduled appointment the leg pain got so bad that he would have gone looking for a surgeon had he not already made the appointment to see me. He agonized over the severity of the pain, recalling the things he had read in my book about the psychological basis for it. Then, out of the blue, he found himself saying something like, "I don't want that job of mine. It's too hard, and there's too much responsibility. I want a job where somebody will tell me what to do."

And, he told me, as this revelation flashed through his mind, the severe leg pain simply . . . disappeared! Over the next few days he continued to have some mild low back pain, so he decided to keep the appointment with me and go through the full therapeutic program. As it turned out, he did well and became totally pain free in a few weeks.

As an attorney might say, I rest my case.

This young man received no treatment, he simply *learned something by reading my book*, and through a mental process managed to eradicate a severe physical pain. Knowledge is power. Let us see what his experience tells us about the psychosomatic process.

FIGURE 1

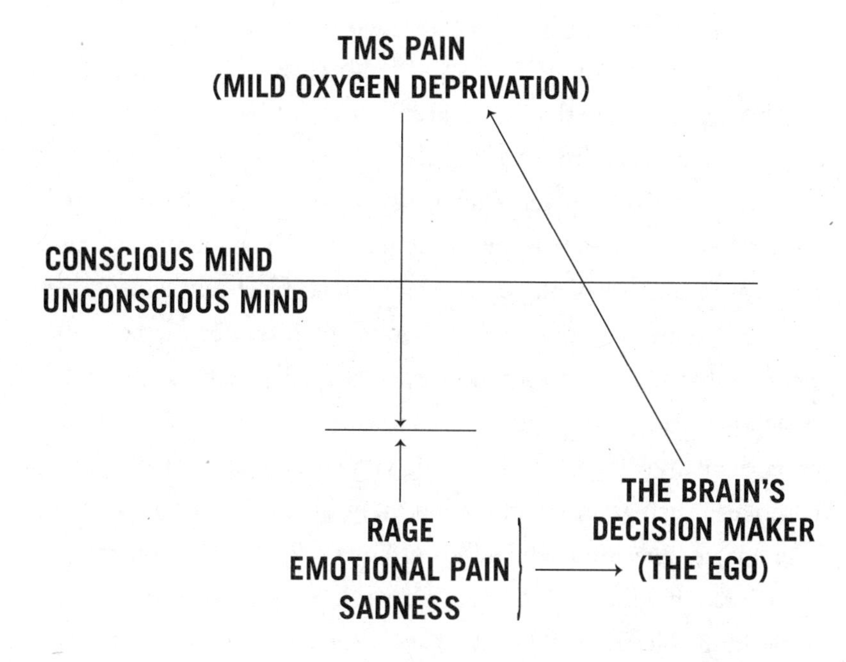

What immediately comes to mind is that this was a very intelligent, well-educated, conscientious, hard-working, high-achieving young man. How did he get that way? His intelligence was probably genetic in its origins. But where did the conscientious, hardworking, success-oriented tendencies come from? There are lots of very intelligent people who never accomplish much, and there are numerous psychological explanations for their lack of success. What then drove this particular young man to do so well that at age twenty-eight he had such an extremely responsible job? And then the most interesting question of all: why did the job give him such pain?

Figure 1 describes the basic physiology of TMS. The figure shows that the physical process resulting in pain is initiated in the unconscious by the brain's decision maker, which Freud called the ego. The ego is aware of the rage, emotional pain, and sadness and concludes it must do something to prevent the explosion of the rage into the consciousness and keep the person from experiencing the pain and sadness. And what does it do? It creates pain as a distraction because it knows that will prevent the rage from coming out and spare the individual from feeling the pain and sadness.

The fact that a psychosomatic symptom cannot be measured with the same precision as fever does not mean it is not real. Its existence, as in the case of the young engineer, is objectively evident. Yet the modern medical community has gone to great efforts to twist the evidence to suit its own preconceived notions. Today, throughout the Western world, virtually every medical practitioner is quick to ascribe the pain suffered by the young engineer to physical-structural pathology, such as an inflammation or, in his case, a herniated disk. There is no question in the minds of any of those practitioners that the source of his pain was truly physical. Yet here is a case where something objectively physical is clearly the result of an unconscious mental/emotional process. One would think that psychoanalysts, who have been criticized for their lack of objective data, would jump at the chance to support such evidence of mindbody disorders, but unfortunately, most psychoanalysts are unaware of the existence of TMS and are unable to capitalize on this important piece of information. That is one of the reasons for this book.

PAIN AND REPRESSION

What is the purpose of the pain? It was Stanley J. Coen, of the Columbia University College of Physicians and Surgeons, who first suggested that psychosomatic physical symptoms were in all likelihood a defense against noxious unconscious emotional phenomena. The concept was presented in a paper published in 1989. Prior to the publication of that paper, I had entertained the idea that symptoms were a substitute for undesirable unconscious feelings. Dr. Coen's insightful perception set me on the conceptual track that led to a true understanding of the nature of psychosomatic processes. I am, accordingly, very much in his debt.

The young engineer's experience made it clear that his pain was a reaction to an unconscious emotion—rage—and its purpose was to assist repression and make certain that the rage did not reach consciousness. Multiple factors contribute to the reservoir of unconscious rage across the spectrum of patients, but the one that brought it to a dangerous level in this patient, the one that convinced his mind's decision maker of the necessity for a physical symptom, was his hatred of his job. And then there was the remarkable phenomenon that occurred when he became consciously aware of that unconscious reaction: the pain went away! How and why that happens will be the subject of chapter 4.

It was one of Freud's early ideas that human beings harbor a great many unacceptable ideas in the unconscious that had to be kept there, leading him to propose the concept of repression to accomplish that end. His early preoccupation was with embarrassing and socially unacceptable sexual thoughts and feelings that he considered to be the primary purpose of repression. My reinterpretations of some of his cases illustrate a different basis for repression.

Psychosomatic symptoms are created to assist the repression of rage and other unacceptable feelings. Although it is not entirely clear why these unconscious feelings strive to become conscious, it is abundantly clear why the brain resists the attempt: some of those feelings are believed to be too dangerous or embarrassing or otherwise unacceptable to be brought into the light of day, while others are simply too painful to be experienced consciously.

Look now at figure 2. If you move down from the top, you encounter the words *rage, sadness,* and *emotional pain.* This is only a

FIGURE 2

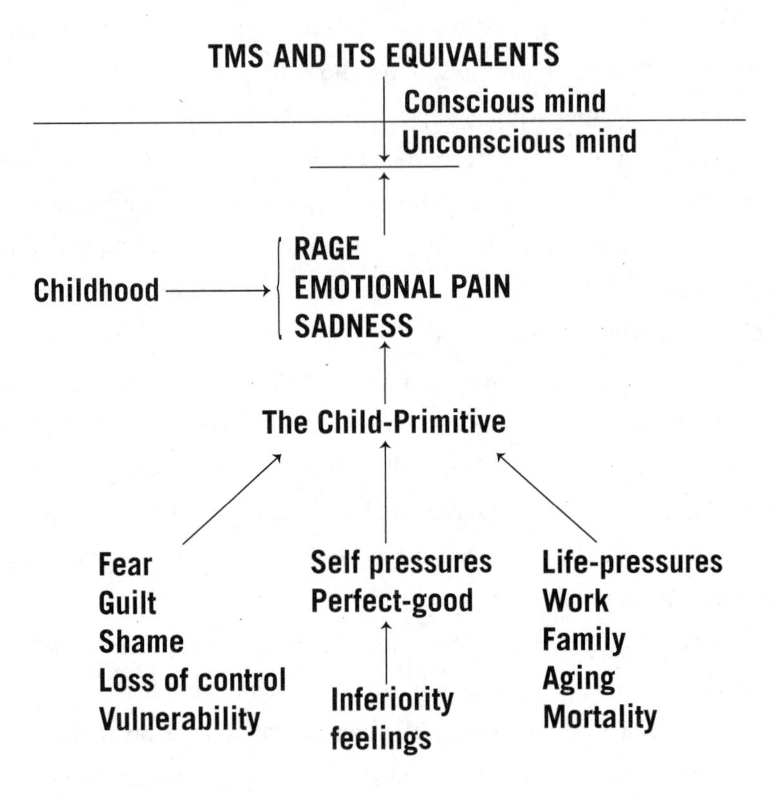

partial list of the emotional phenomena to be found in the unconscious, but it includes the ones of greatest importance—most particularly rage—in the genesis of mindbody disorders. The rage I'm talking about here is an accumulation of anger generated daily over many years and kept repressed for the reason given above. The decision maker in the brain has decided that the overt expression of unbridled rage would ruin the person's life, and to prevent that from happening, it automatically initiates physical symptoms in the body without consulting the conscious, rational mind.

The unconscious rage under discussion must not be confused with conscious anger or rage. While such overt emotions often reflect unconscious rage, they are not the same thing and they are not stimulated by the same phenomena. They are called *displacements* by psychologists. The varieties of road rage are a good example of displaced emotions. In its most benign form the driver finds himself reacting with anger to annoying things his fellow drivers do. He swears at them, berates them, accuses them of gross incompetence, and so on, all from within the safety of his automobile with the windows closed. I'll admit that this is very familiar to me since I do it all the time, much to the distress of my wife when she is in the car. What is particularly striking about it to me is that my reactions are so immediate and strong—no doubt reflecting the intensity of the inner rage for which they are substituting.

This reminds me of a patient who reported that he was so consistently bad tempered that his family urged him to seek psychotherapy. Since he was basically a "good" man, he agreed and after a time learned to control his temper, to everyone's pleasure, including his own. However, not too long after he achieved this desirable state, he began to have back pain for the first time in his life. It grew increasingly severe, and he eventually came to me. Apparently, he needed to express his displaced anger in order to keep his inner rage at an acceptable level. When that option was no longer available to him and the repressed rage increased and threatened to become overt, he developed symptoms—back pain—to keep this from happening. It has been observed in many patients that when conscious anger is *suppressed* it will become part of the reservoir of rage in the unconscious. Does this mean that conscious anger can never be suppressed? Since there are situations that demand suppression (work, family, etc.), one can avoid the negative consequences of suppression by being aware that the anger will not simply go away but will add to the unconscious rage reservoir inside. I had a patient who described a panic attack after suppressing intense anger. Fearing that it would not remain repressed, his psyche arranged an immediate physical reaction as a diversion. It is likely that had he known the consequences of suppression he would not have had the panic attack. This demonstrates just how important it is to be aware of your inner rage.

While it is easy to see why it's important to prevent rage from becoming overt, what about things like *sadness* and *emotional pain* since they, too, may be involved in the psychosomatic process? I am indebted to the psychotherapists who work with me for a rudimentary understanding of these psychodynamics. While the repression of rage protects us from danger, the repression of other intense, bad feelings protects us from feeling emotional pain. It has also been suggested that emotional pain contributes to the internal rage. Here's a seeming paradox: the symptoms created by the brain-mind, which we deplore as evils, are in fact generated for our protection. One of my patients told me, "I think of TMS as a gift." He meant it revealed the true cause of his pain, and it taught him things about himself and his feelings that he would not have known otherwise. In some cases it leads to much needed psychotherapy.

But why should the symptoms take the form of pain, or stomach trouble or an intractable rash or severe allergies, or, for that matter, anxiety or depression? It would appear that all such disorders are meant to distract, to keep one's attention focused on the physical, on the body. Even anxiety and depression, clearly psychological phenomena, are states of discomfort that serve just as well as physical symptoms to distract us from what the brain-mind considers more dangerous or problematic phenomena. The more intense or threatening the unconscious feelings are, the more severe the psychosomatic reaction is likely to be.

Hypochondria is an extreme example of physical reactions to intense unconscious emotions. It defines a state of mind characterized by extreme preoccupation with the body. The patient will have many symptoms and be certain that he or she has serious illnesses. Some of the symptoms may be psychosomatic, some hysterical, some phobic, but classifying the symptoms is of no importance. Hypochondria should be treated with the kind of psychotherapy that explores the workings of the unconscious mind. This is referred to as dynamically or analytically oriented psychotherapy.

Let's take stock. Thus far, our young engineer has taught us three important things about the psychosomatic process:

1. That physical symptoms (in his case, back and leg pains) are a reaction to unconscious mental/emotional phenomena
2. That these unconsciously generated feelings are repressed as a matter of self-preservation
3. That becoming aware of them can lead to "cure"

But to truly understand what happened with the young engineer, we must first look more closely at who he was and how he got that that way.

Refer again to figure 2 on page 94. The sadness, emotional pain, and rage are unconscious reactions to three sets of emotions:

1. Those resulting from the experiences of infancy, childhood, and adolescence
2. Those based on responses of the child-primitive in each of us, that is, to both self-imposed pressures, and the pressures of life

3. Those due to a miscellaneous group of unconscious emotions

It bears repeating that the ideas expressed here are the result of years of experience and thousands of patients treated by the author and psychotherapists who work with him. The accuracy of these ideas can be attested to by the high level of success we have had in working with these patients.

EMOTIONS GENERATED IN INFANCY, CHILDHOOD, AND ADOLESCENCE

To my knowledge, there is no evidence that infants generate anger, or some similar reaction to problematic emotional dynamics, but it would not be surprising if this were to be demonstrated someday. For instance, it is known that the bonding between mother and child in the early months of life is a vitally important, highly sensitive process. But it is not hard to imagine a less-than-optimal situation in which the mother has problems—either practical or psychological—that might interfere with the bonding process, thereby stimulating a negative emotional reaction in the infant.

One suspects that children born into the relatively simple, socially uncomplicated environment of a primitive society, such as may still exist in the wilds of South America, may be psychologically healthier than their counterparts in, let's say, New York City. The infant is carried in a sling as the mother goes about her daily tasks and feeds at the breast when it's hungry. The bonding is bound to be good. (I well recall experience with my daughter, who was a notorious crier and nonsleeper as a baby. She never slept during the day, but when she was on my back in a carrier I could participate in all sorts of activities, even pick-and-shovel work, and she would sleep through it all.) Then later, during the childhood years, the mother or father trains the child of the same sex as it matures and learns its role in society. One should not idealize primitive society, but there is little doubt that in the matter of child rearing it is more straightforward than ours.

The child growing up in the modern world, regardless of socioeconomic status, may be the victim of emotional, physical, or sexual abuse, all unconsciously enraging. The abused child may feel sad, unhappy, or scared, but anger is forbidden—that prohibition is built in. But it is there, in the unconscious, and it stays there. It accumulates and it is permanent because there is no sense of time in the unconscious.

The longer I do this work, the more impressed I am with the ubiquity of a certain kind of emotional abuse of a subtle nature, not always easy to detect. Invariably, when I ask patients to characterize their childhood years they will tell me they were okay, or normal, or even fine. But then, when pressed for details, another picture begins to emerge. When I ask, "What was your father like?" they will admit that he was a stern disciplinarian, or he wasn't around very much, or he left the raising of the children to Mother, or he had a short temper, especially when he had been drinking, or he and Mother were constantly fighting, or they separated when the patient was five or ten or fifteen.

An extremely interesting and crucial fact concerning the emotional development of boys in particular, but to some extent girls as well, is that previous generations of men never considered they had a role to play in bringing up children beyond that of providing for their physical needs and perhaps setting rules of behavior. They were not aware of boys' need for warmth, closeness, total acceptance from the father—that the least important and possibly detrimental role was that of father as disciplinarian. There are many ways of setting rules; the growing child need not be intimidated by a parent and should not be in constant need of avoiding disapproval. Many people carry this need into their adult lives because it was so deeply ingrained in childhood.

"And what of your mother, was she warm and loving?" Well, she was sick a lot and we kids had to do a lot of things for her, or I could never be good enough to suit her ("good little girls don't behave like that"), or she always seemed to be putting me down. If I got ninety-eight on a test, why didn't I get a hundred?

The developing child needs warmth, approval, role models, and guidance. We are beginning to see more of this nurturing capacity in contemporary generations of parents, but it appears that social patterns of the past were often neglectful or hurtful to the developing child, wounding the child's self-regard, with all the ramifications of low self-esteem, and thus creating emotional needs that persist through life. We yearn for that which we did not get as children and are permanently sad, hurt, and angry as a consequence, but all in the unconscious. This is the stuff of which psychosomatic symptoms are made.

When the hurt is gross, as with sexual abuse, it may be the major contributor to the unconscious rage and emotional pain. This was what happened with Helen, whose case we touched on briefly in chapter 2. Her story also vividly documents the permanence of unconscious feelings. Though she had repressed the memory of sexual abuse by her father for many years, when the memory returned, her violent emotional reaction made it seem as though it had all happened the day before.

SELF-IMPOSED PRESSURES: THE PERFECT AND THE GOOD

Figure 2 demonstrates a link that connects *unconscious feelings of inferiority* to the pain of TMS. Feelings of inferiority are no doubt enhanced by a childhood lacking some of the things noted above, which may have been the case with the young engineer, for in order to achieve what he did he must have expected a great deal of himself and put himself under great pressure.

As noted in chapter 2, Freud suggests a more universal reason for feelings of inferiority, namely, that the transition from the childish state to adulthood is painful and distressing and leaves behind "a permanent injury to self regard in the form of a narcissistic scar, which in my opinion . . . contributes more than anything else to the 'sense of inferiority' which is so common in neurotics."

It is my view, as it was Adler's, that feelings of inferiority are *universal*, varying in degree and intensity, but not limited to "neurotics." The sense of inferiority appears to be the primary spur to the drives to be perfect and good that is exceedingly common in patients suffering with TMS. It caused Adler to conceptualize the idea of the *superiority complex*, which is synonymous with what we have described as the drive to be perfect. It must be that we are unconsciously trying to prove to ourselves and the world that we are worthwhile, not inferior. As a patient of mine once put it, success is often built on insecurity, which is another expression of the same idea. Feelings of inferiority are of immense importance if one is to understand the psychosomatic phenomena they generate. The drive to superiority (perfection) is a *pressure*, which brings us to the next link in the chain, the *child-primitive*, Freud's id.

THE CHILD-PRIMITIVE

We touched on the ancient origins and the evolution of Freud's id in chapter 2. The anatomical persistence of the "old brain" described there and all the behavioral characteristics that are, so to speak, contained in it, is of immense importance not only socially, politically, and medically but psychologically as well. It is the part of the brain whose inclinations are in direct conflict with the more responsible, intelligent, moral propensities of the "new brain." The child-primitive is totally narcissistic, irresponsible, and dependent. It harbors the seeds of violence, lust, and obscene behavior and reacts to the pressures of the responsible mind with anger. It cannot stand the pressures to be good, or perfect, or to take care of others, no matter how emotionally close to those others they might be. A perfect example is a young couple with a new baby who happens to be very difficult, crying much of the time. Consciously, the parents are exhausted and worried; unconsciously, they are furious at the baby.

For many years we were aware of a consistent connection between the perfectionist tendency and the development of TMS. Though some of our patients denied being perfectionists, they admitted to being hardworking, conscientious, responsible, driven, success-oriented, perpetual seekers of new challenges, sensitive to criticism, and their own severest critics. But it was a long time before I realized that the drive to be *good* could be equally enraging to the child-primitive. It is yet another pressure. Though both the *perfect* and the *good* tendencies are present in most patients I see, the good is often the predominant one. These people are aware that they have a great need to be liked and so find themselves looking for approval in everything they do and going out of their way to be helpful, often at the expense of their own comfort and convenience. If you speak to someone who is an inveterate caretaker, helper, do-gooder, you are impressed by the power of the compulsion that drives them. Typically, they avoid confrontation.

These tendencies, like so many aspects of the psychosomatic process, seem to be almost universal. In his biography of Freud, Peter Gay said, "Freud thought that the neurotic throws such a clear light on the normal largely because the two are really not so different from each other." Our own findings go considerably further than that. We do not think our patients are neurotic. The psychosomatic reactions they are experiencing are both normal and universal.

The drive to be perfect and good are reactions to *feelings of inferiority*, which are always unconscious (and sometimes conscious as well). Such tendencies to achieve and be nice are typical of people trying to demonstrate by their performance and behavior that they are worthwhile, not inferior. Statistically, these tendencies are among the most important contributors to the reservoir of rage that is a primary factor in the genesis of psychosomatic symptoms.

In a survey of 104 of my TMS patients, I found that the perfect-good drive was either the *predominant factor*, or a *very significant factor* in 94 percent of the cases. Based on life history, I found that:

- In 31.5 percent of the patients studied, the *perfect-good tendency* was the primary contributor to the TMS rage, and that childhood abuse and life pressures were less important.
- In 36.5 percent of the cases, the *perfect-good drive* and *life pressures* appeared to be equally significant contributors.
- In 17 percent, the *perfect-good drive* and *childhood abuse* were the most important.
- In 8 percent, all three factors (*perfect-good drive, life pressures*, and *childhood abuse*) were equally important.
- In 3.5 percent, *child abuse* was the primary contributor.
- In 2.5 percent, *life pressures* were the most important contributor.

Striving to be good is not a sickness, and statistics like these make it clear that psychosomatic disorders and the psychology that prompts them are not pathological or "neurotic" but an integral part of normal living. Yet mainstream medicine continues to consider them pathological for the same old reason—because they have been misdiagnosed as due to structural or soft tissue pathology. When viewed properly, they are clearly part of normal human experience. You can study the anatomy, physiology, and chemistry of the brain forever and a day, but it will teach you nothing about psychosomatic phenomena, nor about emotional disorders like depression for that matter. This is because the deviations from the normal that are featured in such studies are not the cause of the disorders but rather the result of the emotional phenomena that stimulated those deviations in the first place. As I've already said, it is emotions that drive the chemistry in the brain, not the other way around. Altered serotonin chemistry is not a disorder, it is an emotionally induced chemical reaction resulting from the true symptom, which is depression.

We are much further along now in understanding why the high-achieving young engineer developed TMS. He had pushed himself unmercifully, and his child-primitive reacted with rage. From there on, the sequence of events was inevitable—his decision-making ego had to instigate TMS or one of its equivalents to make sure that the rage did not erupt into his conscious life. The fascinating thing in his case was that despite the ego's efforts, the child-primitive mind managed to make itself heard. "I don't want that job—I want a job where someone will tell *me* what to do."

THE PRESSURES OF LIFE

The child-primitive is under pressure not only from our self-imposed drives to be perfect and good, but from the many things going on in our outside lives: our work, career, family (both immediate and extended), finances, illness, aging, and mortality, just to note the most important. These pressures are easier to understand than the self-imposed ones because they can be observed objectively. Taken together they contribute significantly to the reservoir of rage we all have. Traditionally, these environmental *pressures* have been referred to as stressors. We prefer to call them *pressures* because that term includes things not generally thought of as stressors, like being a good wife or husband, or a good parent, or a good son or daughter to one's parents, and because it also carries the idea that something in the psyche is being pressured, an idea that is essential to the concepts advanced here.

Once more, it is essential to note the great disparity between *conscious* and *unconscious* reactions to life events. Similar to the situation of the young parents who are unconsciously furious at their crying baby is one at the other end of life's spectrum, where an elderly parent requires care. No matter how willingly the care is provided consciously, the child-primitive within us will react with anger and resentment; and the nicer the care-giver is, the greater will be the internal reaction and the possibility of psychosomatic symptoms. In cultures where large families are the rule, especially when religious convictions mandate many children, mothers may develop psychosomatic symptoms, unaware of the inner rebellion that has been brought on by the enormous amount of work and responsibility they must endure.

There are all sorts of pressure in the workplace, many quite obvious but some harder to spot. Take the case of the man who employed about fifteen people in his business. Because he had a great need to be a "nice guy," he was constantly worried about whether his employees were being well served by their jobs. This was clearly an important factor in his developing the back pain that brought him to my clinic. He was a great employer for his workers but very bad for himself.

Financial problems and illnesses are obvious examples of life pressures and need no further explanation. They are consciously disturbing and unconsciously enraging. But what about getting old and dying? We tend to rationalize. After all, dying is part of life, it's inevitable, and one must accept it with good grace. But it is a very different story with the child-primitive in the unconscious. That narcissistic part of our emotional makeup is in a rage at the idea of having to put up with illness, perhaps disability, and the ultimate insult to the individual—death. These feelings, though unconscious, are as real as the ones of which we are conscious. With some of my patients the reaction to aging is the only reason for their symptoms. They may be consciously aware of their fear of aging and death but not of their unconscious reactions, and these are the ones that bring on symptoms.

Mainstream medicine is aware that emotions can worsen an existing disease process, but it appears to be unable to understand or accept the idea that symptoms can be *initiated* in response to emotional states. The altered physiology and chemistry that produce these symptoms, whether physical or affective, are induced by unconscious emotions. Thirty years working with mindbody disorders—TMS and a host of others—clearly support this.

THE 9/11 SYNDROME

Many of the emotions that we are aware of consciously are also experienced in the unconscious and contribute to the reservoir of rage, emotional pain, and sadness. The terrifying events of September 11, 2001, resulted in a dramatic increase in psychosomatic reactions across the United States, as might have been expected. The majority of us were frightened, but fright doesn't create psychosomatic symptoms. The reaction to conscious fear is to try to overcome its source, to deny or to rationalize it. But we are unaware of our unconscious fear and the feelings it engenders, feeding our reservoir of emotional pain to such an extent that new symptoms are bound to occur. That is what happened after September 11. There's an important lesson here. If, as Freud theorized, psychosomatic symptoms were meant to be punishment, why would we need to be punished for being afraid? There would be no reason. But if, on the other hand, symptoms are designed to protect us from experiencing painful and dangerous emotions, as we have shown, then our reactions to the terrorist attacks on New York and Washington are logical—the increased bad feelings had to be contained in the unconscious, hence the need for symptoms.

The fear of death and disability, noted above as a life pressure, was certainly enhanced in many people that tragic day.

Guilt and shame are intolerable to the child-primitive and evoke anger, emotional pain and sadness, and there is no dearth of those emotions among all of us. Many firefighters experienced survivor guilt after 9/11, guilt that they did not share their comrades' death. This reaction is well known from the experience of Holocaust survivors.

Guilt is tied in with inferiority feelings and the need to be perfect-good and with all that what those tendencies imply. People blame themselves for lots of things because of their low self-esteem and their perfectionist tendencies. There appears to be a deeply ingrained habit of self-deprecation that is part of the very fabric of their personalities, influencing every moment of their lives. If they had a more robust sense of themselves, they would have a more balanced view of the things they feel guilty about. This, of course, points back to childhood when they didn't get the support they needed to give them this more robust sense of self.

Those of us who are prone to psychosomatic manifestations usually have a strong need to be in complete control of our environments. For obvious reasons, 9/11 challenged that. We lost that sense of control, which in turn led to more internal negative reactions.

Prior to 9/11 people generally believed that they were well looked after, and could depend on the government, the airlines, the police and firefighters to see to their safety, and then all at once they found that they couldn't rely on any of them—and the result was more inner rage, and the potential for more symptoms.

Fear, loss of control, unmet dependency needs, a sense of helplessness, victimization—all such feelings were intensified by the new reality of terror, resulting in frightening and painful feelings in the unconscious and a major increase in psychosomatic symptoms.

Mainstream medicine, hardly aware of psychosomatic phenomena, concluded that the anxiety of 9/11 only made existing disorders worse, which it undoubtedly did, but failed to recognize that many people developed symptoms de novo.

Though triggered by 9/11, these emotions can now be identified in the everyday lives of almost everyone and they continue to make ongoing contributions to our painful internal feelings.

THE IDEA OF UNCONSCIOUS FEELINGS

Generally, people find it difficult to conceptualize the idea of unconscious rage. Some find it abhorrent, while others simply can't believe it can be there inside them without their knowledge. They think anger and rage are such strong emotions that one must be aware of them. The idea that emotions—raw, heated, towering emotions—can exist outside of consciousness is hard to accept. Even when people intellectually acknowledge that these might exist, they find it hard to imagine them because they don't feel them.

We live in the world of the conscious, and most of us think it is our only world. We acknowledge only what we are aware of, what we feel consciously. People exhibiting psychosomatic symptoms have to make an effort to imagine painful or threatening internal feelings and, equally important, reflect on the magnitude of their feelings and their potential for doing great harm. One must learn to think of these unconscious feelings in volcanic terms and understand that their intensity has the potential to wreak havoc in our lives or would simply be too painful to bear.

That is how the decision maker in our brains—the ego—must conceive of the situation, for it stimulates the production of physical or affective symptoms *automatically*, without seeking the approval of the thinking mind. The process totally bypasses the intellect. It is clearly a subcortical reaction, for logic suggests that if reason were permitted to participate in the decision it would likely say, "This is ridiculous. I'd rather deal with the scary feelings than suffer the pain."

But the psychosomatic process does not allow us a choice. The threat to the ego must seem mortal, and the intellect is not permitted to participate in the decision. It is bypassed. The ego acts decisively and swiftly, and induces symptoms. It will not be denied, as illustrated by the following case.

THE SYMPTOM-IMPERATIVE

Mr. O, who was in my cognitive therapeutic program, called and reported that the low back pain for which he had originally come to see me was much better. Then he said, "But I have begun to have pain in my neck and shoulder, my old stomach symptoms have returned, and at times I feel very anxious. It's funny, but when I feel anxious I have no pain."

Here is a striking example of the symptom-imperative, referred to in chapter 1. My clinical program had relieved his low back pain but his psychological state was such that it required continuing symptoms, either physical or affective, so his psyche produced a new location for pain, reintroduced his old gastrointestinal symptoms, and made him intermittently anxious. But one symptom at a time was sufficient to keep his conscious mind preoccupied, so when he was anxious he had neither pain nor gastrointestinal symptoms.

Mr. O readily understood that it was time for some psychotherapy in order for him to explore the unconscious phenomena responsible for this remarkable sequence of events.

Another patient, call him Mr. Q, illustrates a more serious manifestation of the symptom-imperative. He was seen initially for neck, shoulder, and arm pain. He did not improve with the cognitive therapeutic program and was referred for psychotherapy, but quickly found that it was not to his liking. On his own, he decided to undergo physical therapy which provided some relief. About a year later he had cardiac bypass surgery. This was followed some months later by severe shoulder pain. His doctors attributed it to a torn rotator cuff, and on their recommendation he had surgery. Again, he was relieved of pain. Six months later he was found to have prostate cancer and had extensive treatment for that disorder. Many months later I learned from a friend of his that he was now preparing for back surgery.

Mr. Q had multiple manifestations of TMS, but he could not accept a psychosomatic diagnosis. He was a mild-mannered man, always in complete control of his emotions, but someone who felt things very deeply. It is my view that the underlying rage that caused his TMS pains also played a role in bringing on his cardiovascular and neoplastic problems, and that the symptom-imperative was responsible for his multiple pathologies. A recent paper in the *Journal of Psychosomatic Research* makes a convincing case for the role of stress in the development of atherosclerosis. Our experience suggests that *unconscious rage* rather than *stress* is the active psychological ingredient leading to many serious disorders, including atherosclerosis.

The concept of the symptom-imperative is not new. Freud described it a hundred years ago: "What happened was precisely what is always brought up against symptomatic treatment. I had removed one symptom only to have its place taken by another."

Freud believed that the purpose of symptoms was to *punish*, but he did not comment on the reason for the phenomenon of symptom substitution. My experience in treating these cases strongly suggests that psychosomatic symptoms are meant to distract, and to protect the conscious mind from dangerous emotions, and we conclude that the need for new symptoms is to guarantee that the protective mission will continue.

Mr. O's improvement was very real. It came about when he acquired insight into the process causing the symptom. Many of my patients become permanently symptom free simply by coming to understand the nature of the psychosomatic process. But in Mr. O's case that knowledge was enough to dramatically lessen his distress, although not sufficient to reverse the entire process. However, since the sources of his rage were known, the new symptoms merely told us there was still work to be done, and he would have to dig a little deeper with psychotherapy. He could look forward to reversing all of his symptoms since psychotherapy was initiated shortly after his call.

The situation was very different and more dangerous in the case of Mr. Q, because his symptoms were removed by a series of placebos. The symptom was "cured," but the *cause* of the symptom remained untreated. And because there was no knowledge of the true cause of the symptom, the brain simply produced new symptoms, a process that can go on indefinitely, as it does in the case of millions in the United States. The potential danger lies in the fact that the substitute symptom may be due to serious pathology: an autoimmune, cardiovascular, or neoplastic process, as exemplified in the case of Mr. Q.

Another case history, briefly mentioned earlier, beautifully illustrates the workings of the symptom-imperative. The patient was Mr. W, a forty-year-old professional man, married with two young children. About six months prior to his appointment with me, he experienced, for the first time, typical back and leg pain. He had a mild history of irritable bowel syndrome but otherwise had been healthy and very active physically. Imaging studies revealed nothing significant in his spine. Conventional treatment had not been helpful. The physical examination was unremarkable. He had TMS.

His psychosocial history was of great interest. First, he acknowledged that he was a perfectionist and felt compelled to be "a good guy"; then, late in the session as we were discussing unconscious rage, he said he should tell me that some time earlier he had decided to seek psychotherapy because of his violent temper. The process had been quite successful, and he had learned to control his temper. It was shortly after he completed the course of psychotherapy that his back and leg pains began. He further revealed that an older, domineering sister had established his mother in a community geographically far removed from him, making it difficult for him to visit her. This was undoubtedly a source of both conscious and unconscious anger. But now that he no longer had the safety valve of conscious anger-rage, the accumulated rage in his unconscious became threatening in its intensity and the result was physical symptoms in the form of pain in the lower back and leg.

This is the symptom imperative at work. When he learned to curb his temper, the back pains began. *Psychic and physical symptoms are interchangeable, both serving the same psychological purpose.*

Pain management clinics have become common in the United States. The doctors practicing in them are guilty of serious miscalculations. They treat chronic pain as a separate disorder based on "secondary gain," a concept described earlier, that implies that a structural abnormality is the cause of the pain. Since it is my view that the majority of these patients are suffering from TMS, whatever benefits might accrue from their treatments must be due to the placebo effect, and is therefore only temporary, because the symptom-imperative will come into play.

Knowledge of the intricacies and intrigues of the unconscious is Freud's gift to us. But he underestimated the mind's power by failing to recognize that it could achieve its ends by altering physiology through manipulation of the autonomic, immune, and neuroendocrine systems. Years of work with patients suffering from TMS and its equivalents provide abundant evidence of this power.

Who we are and what our lives consist of are the most common causes of psychosomatic symptoms. I recall a young man who realized during a group therapy session that he must have been generating a lot of internal anger because his workplace was an hour and a half from home and he had such a long commute that there was little time left for him to interact with his little girls.

Mr. G, a forty-year-old married skilled worker with a daughter had suffered incapacitating back pain for three years. He described himself as trying to be a model husband and father, as well as a model son to his aging mother. His father had died when he was sixteen, and he had to go to work to help support his mother and a younger brother, and abandon his college plans. Later, he married. He got along passably with his wife, though they occasionally engaged in shouting matches when she did not accede to some of his wishes.

In the course of psychotherapy, Mr. G came to realize that he was unconsciously furious at his entire family—his wife, his daughter, his mother, and his brother. He resented having had to work so hard to support them, and the fact that they had ruined his chance to go to college. He became aware that in addition to the sadness he felt at his father's death, he felt intense anger at having been abandoned by his father and left with such a burden of responsibility.

The existence of these unconscious feelings were difficult for the patient to accept, but as he managed to do so, his pain receded and he found that he was not as prone to manifest conscious anger at what were clearly irrelevant, nonthreatening targets. Displaced anger, consciously experienced, is common when there is significant unconscious rage. It is a safe substitute for expressing the forbidden rage inside. This is undoubtedly the mechanism behind phenomena like road rage.

Mr. G did well in psychotherapy and has remained free of pain for the ten years since he came to see me.

Mrs. B was a fifty-three-year-old married investment firm executive who had suffered back pain for six years. Her income significantly exceeded that of her husband. They had two daughters of college age. She said her marriage was good and had high praise for her girls, who were exceptional students and "good people."

She did not respond to the education part of the program, continued to have pain, and, as is routinely done in such a situation, was referred for psychotherapy. It was only after several months of psychotherapy that she became aware of her anger at her husband for not having been a better breadwinner and for his dependence on her to meet their financial commitments. She was proud of her ability to make a lot of money but resented the fact that what she wanted to do was creative writing and she hadn't been able to pursue that dream because of her responsibilities. She loved her daughters but came to realize that she was enraged at them, too, for the load she carried for their sakes. The self-denial she practiced was enraging to that inner child-primitive. The pain was her defensive response to that rage.

This story had a happy ending. Mrs. B became engaged in psychotherapy with Dr. Arlene Feinblatt, my colleague and coworker for over thirty years, and became pain free as she became aware of her unconscious rage and did something about meeting her own needs. She made time for writing and, indeed, was eventually published. Fifteen years later, she describes only an occasional flare-up of pain when she allows herself to become victimized by overwork.

Mrs. B's case history is an excellent example of the effects of the *perfect* and the *good*. Looking at her family from the outside, one would think everything was fine. Her medical advisors would say the back pain was due to aging processes in the spine. That kind of naiveté in medicine is the fertilizer that nourishes epidemics of pain. *The problem was not in her spine, it was in her life.*

It is likely that Mrs. B would never have engaged in psychotherapy if she had not developed back pain and been admitted to our program. After all, she did not suffer anxiety or depression. She was not mentally ill. Why should she see a psychotherapist? She had a backache, so she came to us. And we were able to convince her that she needed to see a psychotherapist to understand where the backache came from. The practice of psychosomatic medicine identifies a whole new population of people needing psychotherapy, most of whom would never have realized the need but for the pain.

Because almost the entire medical community lacks an awareness of the existence of psychosomatic phenomena, doctors have largely become mere technicians to a body rather than physicians to a person. Mrs. B needed a skilled psychotherapist to improve both her physical and emotional lives. She would never have sought such a therapist under the guidance of her regular physician.

Here are almost verbatim progress reports on five patients from Dr. Robert Evans, another psychotherapist colleague who works with patients with TMS. They are included here simply to show the kinds of emotional states that give rise to TMS. Names and ages have, as usual, been changed.

> Michael is a thirty-six-year-old white married male with a long history of TMS pain. Michael's main issues seem to revolve around dependency on his father, both financial and emotional, with great resentment and rage due to this dependency. Along with this comes great identity confusion and fear of "growing up" and confronting, standing his ground, etc. Self-esteem issues are thus significant as well. Treatment focus will be on all of the above.

> Ana is a thirty-five-year-old Hispanic married woman. She has a history of TMS back pain for approximately ten years. The back pain developed soon after the death of her father, an event which she had not related to the pain. Ana appears to have become the "caretaker" of her mother (an alcoholic) and brother, which causes much rage within, with which Ana is not very much in touch.

> Marilyn is a forty-seven-year-old white married woman. She has had TMS back pain intermittently over the past fifteen years. Our main focus has been on Marilyn's having been abused verbally and physically growing up. Initially, Marilyn was totally out of touch with any emotions related to her childhood. Now Marilyn is beginning to feel and see the rage and emotional pain within, as well as coming to terms with her fear of her father, which causes current episodes.

Another patient was Robert, a thirty-eight-year-old white single man with a history of having undergone back surgery twice, thirteen years and eight years prior to his consultation with me. He had low back pain and was markedly restricted in his physical movements because of fear of inducing greater pain. The interim report from the psychotherapist stated:

> Robert continues to be highly motivated in treatment. His main focus in therapy is on his relationship with his father. Robert is terrified of his father to this day and totally out of touch with any rage that must be there given Robert's intellectual descriptions of past experiences with him. We will continue to help Robert uncover, gently and slowly, the experience of anger while dealing with the current fear of his father.

> Candida is a thirty-five-year-old white married woman with a history of severe TMS of many years' duration. Candida is a classic TMS patient, in terms of the need to be perfect (and the tension that goes with it) in order to feel approved of and loved by parents who were never capable of doing so. We are just beginning to touch on some of the rage.

These are just a sampling of the thousands of cases we have encountered over the years. Psychosomatic disorders are common and universal, and we are all likely to experience them from time to time. One is reminded of the Irish novelist Thomas Flanagan's observation, "We possess ideas, but we are possessed by feelings. They lie too deep for understanding, astir with their own secret life and carrying us with them."

A PSYCHOSOMATIC TRIAD

It occurred to me as I reviewed the case of a patient that there are three powerful unconscious realities that often work together to produce a psychosomatic episode. They are:

1. Deep feelings of inferiority

2. Narcissism

3. Strong dependency needs

Each of these leads to unconscious anger/rage and emotional pain. Let us see how these come about.

In this case, one of my patients, Mr. K, was entertaining his two grown sons in his home simultaneously, one from a previous marriage. The fact that they were in his home at the same time was a coincidence. He began to have symptoms a few weeks prior to their arrival, indicating that he was anticipating their arrival with trepidation. When they came, he put himself at their disposal to be the good father and the perfect host. Let's see how each of these traits contributed to the rage.

In response to his unconscious sense of inferiority, as we have already established, he was strongly driven to be a "good guy." He once asked me, "Why do I get pain when I'm trying to be a nice guy?" Being also very narcissistic, his unconscious reaction to putting himself out (being a good guy) was outrage. In addition, he felt a great sense of responsibility for these sons, and their presence heightened that sense, unconsciously inducing further rage because of the increased pressure on his narcissistic self. In the meantime, there was that dependent streak in his unconscious saying, "This is all wrong—instead of my worrying about them, they and everybody else should be taking care of me, and the fact that they are not is painful and infuriating." The magnitude of these reactions was sufficient to induce back pain.

We are all two people, Jekyll and Hyde. The case of Mr. K is a good example of the two minds at work; the conscious one behaving in the reasonable, acceptable way; the unconscious one saying, "I don't give a damn about anyone or anything else, just leave me alone, except to take care of me." Not very flattering, but there it is.

It is appropriate to designate these three different traits as a triad because they often work together to increase the unconscious rage strongly enough to stimulate the brain to bring on symptoms. Actually, the patient reported that during this period he not only had physical symptoms but was often very anxious, to the point where he had to take a tranquilizer. The anxiety, too, was a reaction to the internal pain and anger.

Here is a similar case history, this time of a famous man whose name and profession cannot be used. We will call him Mr. Jones. He is fifty-eight, married, with two young adult children. He had his first coronary at age forty-two. He is a successful businessman who functions at a very high level, and has been CEO of three major corporations over the years. During his tenure with the last of these, he supervised a very stressful reorganization of the company. When he was fifty-seven, he accepted the leadership of another troubled corporation and threw himself into the job of rescuing it. He began to experience angina and was found to have significant narrowing of the coronary arteries. He had bypass surgery, and only a few months later experienced angina again and went through another cardiac procedure.

Mr. Jones denies that stress has anything to do with his recurrent cardiac trouble. His doctors are silent on the subject.

This is a common medical scenario in the United States. The problem is that both doctors and their patients are unwilling to consider that pressure of any kind can lead to coronary arteriosclerosis because they do not believe that the brain can induce physical pathology. (Freud didn't believe it either.) They are unaware that pressure generates rage in the unconscious and that deep-seated rage can contribute to physiologic alteration—including coronary arteriosclerosis. Indeed, rage may be the major factor in its genesis. They are unaware that the enraged unconscious reaction to pressure is very different from the conscious reaction. The well-controlled man remains calm and copes well under extreme stress, but beyond his awareness the pressure induces unlimited fury in the unconscious. That idea would embarrass him if he knew about it, and when such a possibility is suggested, he denies it vehemently. Almost always, neither doctor nor patient is aware that self-imposed pressure, which is characteristic of ambitious, accomplished men like Mr. Jones, is possibly more enraging internally than the external pressure we call "stress."

It is quite incredible to me that cardiologists ignore the experience of the highly respected Dean Ornish, of the University of California's School of Medicine, who has made it quite clear that coronary arteriosclerosis is linked to emotional factors. Unfortunately, such unwillingness to accept objective evidence is all too easy to understand in the light of medicine's psychosomatic blind spot.

Having reviewed some typical cases of TMS, let's look at two of Freud's famous cases and contrast his interpretation of the meaning of physical symptoms with that of psychosomatic psychology.

The first case is that of the previously mentioned Elisabeth von R, described by Freud in *Studies on Hysteria,* published in 1895. Elisabeth von R was a twenty-four-year-old single woman with a two-year history of great pain and fatigue brought on by walking or standing in one place. When she walked, her trunk was tilted forward but her gait was normal, which is to say it was not pathological. The major location of her pain was the right anterior thigh. Freud found that the skin and muscles of most of both legs were quite sensitive to pressure, particularly over the thighs. His neurologic examination was normal. He concluded that her symptoms were hysterical for reasons which he elaborated at length, as was his practice.

She was very intelligent, tended to be judgmental, and because she was a girl and was expected to marry, she had been barred from pursuing a career, which she regretted. She was proud of her father and had a close relationship with him, and was inclined to put her family's needs before her own.

Her symptoms began mildly over an eighteen-month period during which she nursed her father through a serious cardiac disorder that eventually led to his death. His illness and death were a blow to the family's happiness, which was intensified by a worsening of her mother's health and the marriage of one of her sisters to a difficult man. A second sister married more felicitously, but about the same time the mother's need for eye surgery created a new crisis, and once more Elisabeth was cast into the role of twenty-four-hour nurse.

It was during a summer vacation with her mother and her sisters' families that her symptoms became more severe. Freud was able to discover without Elisabeth's being aware of it, that she was in love with the husband of one of her sisters, and that she envied the sister and unconsciously wanted the husband for herself. The vacation threw her into close contact with him and intensified her unconscious reactions and, therefore, her physical symptoms.

Important elements of Elisabeth's personality are clear from Freud's descriptions. She sacrificed herself for the needs of her father and mother. She had the highest moral standards, and the idea that she was unconsciously in love with her sister's husband was, on a conscious level, totally unacceptable to her. When Freud suggested it, she raged at him.

Freud believed that Elisabeth's mental pain was converted into physical pain. He wrote that the physical pain had an organic basis and was being *used* by the mind to serve a psychological purpose. (In medical parlance, "organic" means that the pain is the result of a purely physical process and, therefore, is not psychosomatic.) He said the pain was characteristic of something known in his day as "muscular rheumatism," where the muscles were painful and sensitive to pressure, and that the psyche used the physical pain for its own purpose, that is, to convert a psychic process into a physical one. Freud also maintained that the location of her pain had symbolic meaning, and that the pain in her thigh could be attributed to the fact that her father's leg rested on her thigh when she treated him.

Viewed in terms of our psychosomatic concepts, the picture looks quite different. As I interpret her case, Elisabeth had the classic characteristics of the perfectionist-goodist. She was the quintessential caretaker and denied her own needs repeatedly.

At one point while Elisabeth was caring for her father a budding romance was interrupted by her need to take care of him. She had fallen in love with a young man and on a particular evening, a rare opportunity for her, she was with the young man and felt very "warm" toward him. On arriving home she found that her father had taken a turn for the worse, and she blamed herself for having left him for her own enjoyment. Following this incident, she had little occasion to meet the young man. Circumstances then took him away, and she believed she had lost her only chance for marriage.

This is a perfect example of the self-denying goodist, in this case one who lost a potential lover and husband. Elisabeth developed pains after this incident. Freud recognized there was a connection, and this is his interpretation:

> It was, therefore, in this relationship and in the scene described above in which it culminated that I could look for the causes of her first hysterical pains. The contrast between the blissful feelings she had allowed herself to enjoy on that occasion and the worsening of her father's state which had met her on her return home constituted

> a conflict, a situation of incompatibility. The outcome of this conflict was that the erotic idea was repressed from association and the affect attaching to that idea was used to intensify or revive a physical pain which was present simultaneously or shortly before. Thus it was an instance of the mechanism of conversion for the purpose of defense, which I have described in detail elsewhere.

My understanding differs significantly from Freud's. As I see it, Elisabeth's unstinting self-abnegation engendered monumental rage and emotional pain unconsciously, and there was no way that these feelings could be allowed expression in view of her family situation, the mores of the time and her own powerful goodist tendencies. The pain came into being as a psychosomatic process—what we call TMS—in order to prevent the overt expression of this rage. If this patient had experienced a real conversion hysterical symptom rather than a psychosomatic one, it would have made no difference. The purpose of the symptom is a defense, but it is defending against the painful and dangerous feelings and the possibility that they may be experienced consciously, an eventuality that the ego could not permit under any circumstance.

Freud's attaching importance to "an erotic idea" is consistent with his view at that point in his career that sexual factors were at the root of many neuroses. This was compatible with the mores of the time and may also have been influenced by his own self-analysis. It was not eroticism that was the primary psychic factor in this case but the rage and pain stemming from self-deprivation and loss.

Elisabeth sacrificed herself to care for her mother as well. Her moral standards were high; she demanded perfection of herself. And she felt unconsciously that fate had done badly by her: no romance, no marital bliss, no children, and no career to compensate for these deprivations.

The result of all this was strong unconscious feelings, and her painful symptoms were designed to prevent them from becoming overt, which she would have considered horrible in the extreme. Freud thought that her pain symptoms were organic—truly physical—as indicated by the fact that her muscles hurt when they were pressed or pinched during examination. What he didn't know was that the brain had induced the "organic" process, producing mild local oxygen deprivation as described in chapter 1, and that the pain was, therefore, psychogenic, not organic—a direct result of her psychological state.

There are a number of issues illustrated by Elisabeth von R's case. Perhaps the most important is the fact that unconscious feelings are the force behind most psychogenic processes, hysterical or psychosomatic. Such a notion rather turns things around. Whereas Freud viewed the suffering of the neurotic, whether psychologically or physically, as punishment, it is now clear that the symptoms are serving to *protect* the individual rather than *hurt* him or her by allowing unwelcome feelings to escape into consciousness. To remind ourselves again, the psyche views the physical symptom as the lesser of two evils. Better to suffer pain than to have one's life ruined by the manifestations of unbridled rage or intensely painful feelings.

The second of Freud's cases is that of Dora, published in 1905. Some of the details of this famous, somewhat controversial case demonstrate that he considered all physical symptoms to be hysterical, including those he identified as "organic" rather than psychosomatic. Since we have stated that all symptoms, hysterical, psychosomatic, or affective, appear to serve the same psychological purpose, the physical difference between a hysterical and a psychosomatic symptom is not an important factor to us diagnostically. It is, however, very important in the history of psychosomatic medicine, in that Freud did not recognize that the mind-brain had the capacity to initiate symptoms. Freud's patient Dora had manifestations of all three types of symptom. Her nervous cough and her migraines were psychosomatic, her aphonia (loss of voice) was hysterical, and her manifestations of depression, irrational hostility, and suicidal thoughts were affective.

At a deeper level, there is some similarity between Dora's case and that of Elisabeth von R. In both cases symptoms were attributed to sexual phenomena but can be ascribed alternatively to rage. In his biography of Freud, Peter Gay suggests as much: "He [Freud] refused to recognize her [Dora's] needs as an adolescent for trustworthy guidance in a cruelly self-serving adult world—for someone to value her shock at the transformation of an intimate friend into an ardent suitor, to appreciate her indignation at this coarse violation of her trust." This we view as another example of one of Freud's errors: that he was not aware of the ubiquity and power of repressed rage.

A personal example would not be amiss here, for it bears the dual stamp of authenticity and accuracy. I was on a trip with my wife when I began to experience severe gastroesophageal reflux. She and I both recognized this to be psychosomatic, and we tried to figure out what was making me unconsciously angry. Was it because:

- I no longer liked to travel because of the inconvenience and discomfort?
- I found some of the places we were visiting disagreeable?
- I would rather be at home working on a book?
- The trip was too long?

We obviously didn't hit on the right answer, because my symptoms continued unabated for the entire trip. It wasn't until we got home that I realized what had been going on. I had promised the long trip to my wife, who loves to travel. I was being a good guy. I was unconsciously furious for having to do something I really didn't want to do. My psyche wouldn't permit me to be consciously furious at my wife, and neither would my reasonable self—so to be absolutely sure the rage remained unconscious, the brain dished up the severe gastrointestinal symptoms. The rage, of course, was the reaction of that unconscious child-primitive—selfish, narcissistic, and totally unconcerned with the needs and desires of anyone else. Not very flattering. Earlier in this chapter I suggested that reasonable people might prefer to deal with the unconscious rage rather than the pain and discomfort of a psychosomatic symptom, if the brain's decision maker would only give them the opportunity to choose. My gastroesophageal reflux episode prompts me to rethink my position. Rage at my wife would have been inappropriate and unfair—better to suffer gastroesophageal reflux!

Viewed from the standpoint of the brain's strategy, one must conclude that the drive to produce symptoms to divert attention from unconscious feelings is so compelling that the brain will stoop to trickery. It will often initiate symptoms in association with essentially benign physical maneuvers, things that the person has done thousands of times before without ill affect, so that the patient will conclude that the physical act initiated the pain. Despite the lack of logic, people are very prone to make such assumptions. They do not know that the brain has tricked them, and even if they are told so, most will not believe it. The phenomenon of "tennis elbow," which we touched on in chapter 1, is perhaps the most familiar example of this, but there are many others. For instance, a golfer executes a swing of the club precisely as he has done thousands of times before. He feels immediate pain in a shoulder or elbow, upper or lower back. The golfer's conclusion is that he has hurt himself, which is exactly what the brain wishes him to conclude because it has decided that the psychological situation demands psychosomatic symptoms at that time. Does that sound fantastic? I have seen it played out in thousands of similar situations, often when the physical act is ridiculously benign. Unfortunately, the deception is usually aided by the patient's medical advisors because they, too, are unaware of the psychosomatic process.

The brain practices many similar deceits. Following a legitimate injury, such as a sprained ankle, the brain will sometimes continue generating the pain in the form of TMS long after healing has occurred. It will often locate pain at the site of an old injury, such as a previous fracture, for example. If it knows that there is a disk abnormality, it will initiate pain somewhere in the vicinity of the disk bulge or protrusion. It usually does so in a somewhat inefficient way so that the diagnostician can discern that the abnormality does not explain the location of the patient's symptoms. This is very common.

Clearly, the brain considers unconscious feelings to be infinitely more dangerous or painful than pain, or why would it practice so sedulously to deceive?

One of the prime characteristics of TMS is that the pattern of symptoms will develop as a result of Pavlovian conditioning. People will experience the kind of symptoms they have learned to *expect* to experience, just as Pavlov's dogs learned to associate the presentation of food with the ringing of a bell. Elisabeth von R had pain associated with standing and walking, though there was nothing neurologically wrong with her. Another patient with similar pain will say that it is sitting that brings on the pain, while walking relieves it. Experience with large numbers of patients at our clinic makes it clear that these are programmed responses that bear no relationship to anything beyond what the patient is conditioned to expect. That's why the most common psychosomatic disorders are invariably the ones currently in vogue.

As noted in my brief sketch of Franz Alexander's work, he and a number of other investigators who published in the first half of the twentieth century identified many psychosomatic disorders, though they had no idea that the pain syndromes common to them all were also psychosomatic. However, soon after Alexander passed from the scene, Western medicine lost interest in psychosomatic medicine. The awareness of the oneness of mind and body, which Alexander thought in 1950 was about to become the dominant philosophy in medicine, vanished with hardly a trace. Today, virtually no one in medicine, including many psychiatrists, believes psychosomatic processes even exist. As a result, epidemics of pain, depression, and life-threatening conditions flourish throughout the industrialized world.

FOUR

TREATMENT

I remember the time I was explaining mindbody medicine to a particularly intelligent but somewhat skeptical patient. At one point she looked at me with the faintest hint of a smile and a wry twinkle in her eye. "And the pain is supposed to go away?" she asked. "Just like that? I'd like to believe you, Dr. Sarno, but I'm not sure I can manage that great a leap of faith."

She didn't understand. "What you need in order to get better, my friend," I said, "is not a leap of faith but a leap of understanding."

That is the heart of treating psychosomatic disorders.

In the early 1970s I began to suspect that TMS was psychosomatic because almost all the people I was seeing had a history of one or more psychosomatic manifestations in the course of their lives. Although the idea seemed absurd that this very physical complaint, back pain, could also be psychosomatic, my own medical history and personality as well as nine years as a family physician made me receptive to the idea. So I put it to the test, and almost immediately I began to see positive results for the first time since being exposed to people with these symptoms. For years I had made the conventional diagnoses: back sprain or strain, muscle pull, weak abdominal muscles, one leg shorter than the other, various spinal disorders, narrowed disk spaces, even sciatica if there was leg pain, though it wasn't clear how the sciatic nerve could be involved. Those remarkable diagnostic tools, computed tomography (CT scan) and magnetic resonance imaging (MRI), had not yet been developed, so one had to do a myelogram (an x-ray with dye introduced into the spinal canal) to diagnose disk herniation, but even that test wasn't always definitive.

Standard nonsurgical treatment included bed rest in the early stages; medications such as "muscle relaxants" and analgesics; physical therapy, consisting of deep heat, most commonly delivered by ultrasound; deep massage; active exercise; and injections of analgesics or steroids. Therapists were also trained to issue a host of admonitions and prohibitions about posture, how to bend and lift, and the like. Physical activity was sharply curtailed, and it was not unusual for patients to be told that there were certain physical pursuits that must be given up permanently.

The results of such treatment were poor and unpredictable; my practice was frustrating and unfulfilling. Failure was inevitable because I didn't know what I was dealing with. The point is so obvious that it almost seems unnecessary to restate it, but *successful treatment depends on an accurate diagnosis*. My colleagues and I have been successful because we have made the right diagnosis, not because we have found the right treatment. We do not have an "approach" to the problems of acute and chronic pain—we have the diagnosis.

The treatment that emerges with any malady is a natural consequence of the nature of the disorder. So it has been with TMS and other psychosomatic disorders.

It soon became evident that *knowledge* was the key to treating TMS. Whatever else might be necessary to effect total resolution of the patient's problem—psychotherapy, for example—people needed to know about the anatomy, physiology, and psychology of TMS and how these were related. This was made particularly clear to me when I encountered patients who were practicing psychoanalysts, men and women who were thoroughly familiar with their own personalities, unconscious dynamics, and life issues—all the things we have cited as important in the genesis of TMS—but who nevertheless had the disorder because they knew nothing about TMS.

Another crucial therapeutic element became clear early on as well: the person must not only *understand* the nature of the process but be able to fully *accept* it as well. Not faith, but acceptance of the idea is essential. Blind faith leads to a placebo cure, if any. By contrast, acceptance and acknowledgment produce permanent results. Failure of acceptance is an impediment to "cure" for some patients because inability to accept the concepts of TMS is one of the psyche's strategies for maintaining the process. As put succinctly by a young woman patient years ago, "Denial of the syndrome is part of the syndrome." In addition to creating pain, the psyche creates doubt—the better to keep the syndrome going.

This was clear in the early days since only the patients who understood and accepted got better, despite the fact that my concept of the psychology at work was then rudimentary. Even at that time, however, there was awareness of the fact that people who were conscientious, hardworking, very responsible, often perfectionist, seemed prone to develop the disorder. It wasn't until years later that the "goodist" tendency was recognized as equally important.

Here, then, was another absurdity. Not only could physical pain be psychosomatic, but you could stop it by learning about it! Quite incredible, and to this day I find it hard to believe. It's almost too good to be true.

But we can add even a third absurdity: Large numbers of people have "cured" themselves by reading one of my three books on TMS. This includes the first, *Mind Over Back Pain*, despite my lack of knowledge of the psychology at the time. The second, *Healing Back Pain*, has been the most successful in that respect. The following letter is exemplary:

> Dear Dr. Sarno,
>
> I want to personally thank you for writing *Healing Back Pain: The Mind-Body Connection*. Your incredible book enlightened me, inspired me, and challenged me to quite literally change the course of my life. I am truly and eternally grateful to you for sharing your knowledge and vision.
>
> For twenty-four long years, beginning in 1976, I suffered through periodic episodes of intense and physically debilitating back and neck pain. With each passing year, my flare-ups became increasingly frequent, more physically debilitating, and of longer duration. In search of relief, I visited countless physicians and chiropractors over the years.
>
> By 1998, my condition had worsened to the point where I was sometimes forced to crawl due to the pain and spasms. At that time, MRIs revealed two herniated lumbar vertebrae and two partially crushed cervical vertebrae.
>
> Over the next two years, I would undergo literally hundreds of hours of treatment. The treatments included physical therapy, rolfing, acupuncture, spinal injections of steroids, consultations with various surgeons, as well as consuming many different types

of muscle relaxers, anti-inflammatories, and other prescription medications. The end result of all of these treatments: I wasn't able to sweep the kitchen floor without ending up physically contorted and in severe pain for days.

I was locked into a slow, progressively downward spiral toward an existence in which my life would ultimately revolve almost entirely around my back pain. In desperation, and with considerable skepticism, I finally decided to read *Healing Back Pain*. My personality traits leapt up at me from off of every page. Overly conscientious, exceedingly responsible, compulsive, my need to have the behavior of others conform to my own narrowly defined standards, and above all else, my uncompromising perfectionism.

Upon completing *Healing Back Pain,* I absolutely knew beyond any doubt that I was about to shed the psychological chains that had shackled me physically for twenty-four long years. It was with a sense of excitement and a single-mindedness of purpose that I dove headlong into my recovery. I followed the recommended treatment strategies to the letter. Within four weeks of beginning my recovery, the pain in my back was completely gone. It will be three full years this coming June and I have never once looked back. Had I never read *Healing Back Pain*, I quite honestly don't know where in my life I would be right now.

My most sincere thanks.

Mr. M

The fact that my books can "cure" is convincing evidence that our therapeutic program is not a placebo. There is no treatment, no charismatic personal influence—only the acquisition of information, which cannot in any way be construed as a placebo treatment. Blind faith is not involved, and, as the letter makes clear, the result is permanent, which is not at all characteristic of the placebo effect where return of symptoms or symptom substitution is the rule because of the symptom imperative.

Because acceptance of the diagnosis is essential for a positive outcome and because so few people are open to such a diagnosis, I have a telephone conversation with all who call for an appointment. After years of experience it is not difficult to determine whether someone is a good candidate for the program, and for those who are not it is a kindness to them and to me to discourage them from making an appointment. This is not discriminatory but simply faces the reality of what is required for successful treatment. It is analogous to a surgeon's decision not to operate on someone who is not a good surgical risk. There is no doubt that this practice has been a factor in our statistically successful treatment program. The screening also enables me to identify those who are suffering from a disorder that is clearly not psychosomatic.

THE PROGRAM

It begins with a consultation, taking a history, a physical examination, and postexamination discussion of how the diagnosis pertains to the patient. One learns first the details of the pain syndrome, a description of onset and subsequent course, the precise locus of the pain, its pattern, and, particularly, its impact on the person's daily life. The past medical history, medications prescribed, and diagnostic studies are recorded.

Of primary importance to me is each patient's psychosocial history: marital status; children, if married; place of birth and the quality of life during childhood and adolescence, with particular reference to the personalities of parents and relationships with them; the ages and health status of parents now; relationships with siblings; and education and occupation history. Patients are asked to comment on how they see themselves as personalities; to list the stressors in their lives, and whether or not they have been in psychotherapy.

The physical examination establishes the doctor's bona fides to be dealing with a physical disorder, of great importance with psychosomatic problems since most patients will reject a psychosomatic diagnosis made by a psychologist, psychiatrist, or social worker. It also provides an opportunity to point out the meaning of symptom location, the details of neuroanatomy that help to clarify the basis of the pain or the inconsistencies of a structural diagnosis, and many other relevant physical factors. The patient's education begins during the physical examination.

If the diagnosis is confirmed by the physical examination, the remainder of the consultation is devoted to discussing the process and how it pertains to the patient. The details of the treatment program are described, beginning with the basic two-hour lecture.

THE BASIC LECTURE

TMS is characterized by a large variety of symptom complexes, but since they all serve the same psychological purpose and there is great similarity in the underlying psychology, it is appropriate to bring patients together for educational purposes. The lecture was instituted many years ago when it became apparent that *knowledge of the disorder was essential to successful treatment*. Over the years it has become clear that the lecture's therapeutic power is firmly based on the information transmitted. We know this from patient reactions during the lecture and from the fact that it is often followed fairly quickly by pain relief.

Before the lecture is begun it is pointed out, after the roll is called, that attendees are indeed in school, that they are there to learn because experience has made it clear that knowledge is the ultimate component for recovery from TMS. In order to get the brain to stop the psychosomatic process, they must:

- Repudiate the physical-structural explanation for the pain and attribute it instead to the benign altered physiology, the physical-emotional basis of TMS
- Recognize that the pain is a reaction to a psychological state and that the tendency to have the physical reactions of TMS and its equivalents is universal and a normal component of everyday life

It is essential to establish the fact that nothing will change symptomatically unless those two conditions are met. This follows from the fact that the purpose of symptoms is to divert attention from unconscious emotional phenomena. If patients denigrate the physical symptom and focus instead on things psychological, they have effectively undermined the unconscious brain's strategy. This is not mere theory—it has been demonstrated in thousands of patients. The lectures are designed to help that cognitive transformation. The acquired information apparently induces a positive reaction in the unconscious since the unconsciously generated symptoms cease. I shall speculate later on what that positive reaction might be. Currently, the lecture consists of two parts, the first describing the anatomy and physiology of TMS as well as the multiplicity of structural conditions, such as intervertebral disk pathology, to which symptoms are routinely and mistakenly attributed. The second part takes up the psychology and treatment of the disorder.

The lecture begins with a review of the history of musculoskeletal disorders over the last fifty-odd years, of the epidemics that have occurred as described in chapter 1. Patients need to know that epidemics of psychosomatic disorders will occur if the condition has been attributed to "structural-physical" causes, if the disorder is in vogue, and if treatment is available, as demonstrated by the Norwegian experience.

Regarding the physiology of TMS, it is pointed out that although pain and neurological symptoms like numbness, tingling, and even weakness are common and sometimes severe, the mild oxygen deprivation responsible for them is benign and does not cause damage to the muscles, nerves, or tendons that may be involved, and the symptoms are transient. This is important since patients almost invariably have been misinformed and frightened by their advisors, medical and nonmedical, whose advice tends to worsen the symptoms. If the patient is unfortunate enough to have some muscle weakness, he/she has often been advised to have surgery in order to prevent "permanent nerve damage." That's because the medical community, which fails to recognize TMS, therefore assumes—without objective evidence, incidentally—that if a structural abnormality shows up on imaging studies, it must have damaged a nerve.

Many years ago I concluded that the structural abnormality was not the true cause of symptoms in most cases, because I found no correlation between the location of the pain and muscle weakness, if such existed, and the structural aberration. As noted in chapter 1, it is remarkable that practitioners can undertake a serious treatment like surgery in light of such blatant clinical inconsistencies. One of the dirty little secrets of the medical community is that clinical medicine is clearly not always as scientific as many would like to think. Of the thousands of patients with disc pathology that we have seen over the last thirty-two years, *not one* has ever developed "permanent nerve damage" after refusing surgery and following treatment for TMS.

Since it is important for patients to know what is *not* causing symptoms as well as what is, the lecture also notes the most common structural abnormalities mistakenly blamed for pain. In addition to disk pathology, it includes such entities as degenerative changes in the spine, spinal stenosis, spondylolysis, spondylolisthesis, and scoliosis.

Included in the physical portion of the lecture is mention of the most common equivalents of TMS, such as the ubiquitous upper and lower gastrointestinal disorders, dermatologic and allergic conditions, common headaches, dizziness, and tinnitus, all of which are commonly psychosomatic when more serious disorders have been ruled out. Patients need to know that TMS, though it is currently the most common of a large group of psychosomatic disorders, is not unique.

Following this, there is a description of the muscle, nerve, and tendon involvement typical of TMS. While all the subjects included in the lecture are covered in my three books, presenting them verbally has a greater impact than the written word.

A very important characteristic of the syndrome is that patients will develop a symptom pattern very early in the game based on Pavlovian conditioning. They will recite in great detail when and under what circumstances they experience pain and assume that there is something problematic about a particular activity or bodily position or time of the day or night that seems to initiate or aggravate the pain. It has become clear that these patterns are programmed a la Pavlov for they disappear promptly with treatment, which would not be the case if they were due to a structural abnormality. Patients are reassured by the knowledge, for example, that the sitting or walking that inevitably brings on the pain is not bad for their backs.

The first part of the lecture is concluded with a discussion of the important role of fear in the perpetuation and intensification of the syndrome. Patients have been frightened by words like *degeneration, disintegration*, and *herniation*, used to describe their structural abnormalities; they have become fearful of physical activity and warned against assuming certain postures ("never bend at the waist without bending your knees"); if pain has been persistent, they fear the consequences of continuing pain and disability, worry about their work and responsibilities, and so on. The deleterious influence of fear needs no explanation. Patients need to be reassured that they have been misinformed and that they are experiencing an essentially benign disorder.

The second part of the lecture covers the material in this chapter and chapter 3. We start by describing what might be called the basic constituency of the unconscious, namely, the id, the ego, and the superego. Then we examine the troublesome, negative, threatening inhabitants of the unconscious like its childlike narcissism, dependency, and feelings of inferiority. Finally, we focus on the unconscious rage, emotional pain, and sadness that are directly responsible for the physical symptoms. I emphasize that it serves little purpose to concentrate solely on the rage itself, since except under extremely unusual circumstances it will not be expressed consciously. I tell the patients that they must, instead, pay attention to the four sources of the rage and emotional reactions:

- Anger, emotional pain, and sadness that can be traced back to childhood
- Anger stemming from the self-imposed pressures to be perfect and good
- Anger generated from the pressures of life
- Miscellaneous things like guilt, shame, fear, insecurity, and vulnerability that also feed the reservoir of rage

Though it is likely that the major purpose of the pain is to prevent the escape of rage into consciousness, we emphasize that the psyche is also desirous of preventing the person from feeling the emotional pain and sadness that are such common legacies of childhood.

Throughout the discussion of the sources of rage, patients are reminded of the great difference that exists between the conscious and unconscious reactions to life, differences that are reflected in the title of this book. We humans are anatomically and behaviorally two different people, in constant conflict.

Helen's story, touched on earlier, dramatically illustrates the perpetuity of painful childhood experiences and the purpose of physical symptoms in the psychosomatic process. It's an important part of the presentation. I had seen and treated her successfully for back pain about a year prior to receiving a letter from her. She wrote that she had begun to remember, after total amnesia for the experience for decades, that she had been sexually abused by her father. She decided to attend a meeting of victims of incest in her community, which was of course a very emotional experience, and later that day she began to have low back pain. She reassured herself that it was TMS and that there was no cause for concern, but over the next thirty-six hours the pain got progressively worse. By the morning of the second day she was virtually paralyzed with severe pain and could not understand why it had gotten so bad in light of her knowledge of TMS. What she didn't know was that during those thirty-six hours the poison of her monumental rage, shame, and pain, which had lain dormant all those years, and which had been stimulated by the meeting, was forcing its way closer and closer to consciousness. The psyche, in a desperate attempt to prevent the explosion of those feelings into consciousness, was making the pain worse and worse. And then the psyche lost. Helen began to cry as she had never cried before, she raged, she wanted to cut her wrists and die. The poison poured out of her and, as it did, virtually all the pain disappeared.

I then tell my patients that their rage will not come out as it did in Helen's case, for the provocation in her case was exceptional. What her case makes crystal clear is just what the role of pain and other symptoms is in psychosomatic disorders. It is there to prevent feelings from coming out. Her experience was the exception that proved the rule; an idea that I had never understood before. The rule: your rage will not come out. The exception: it may if the provocation is great enough.

It is suggested at the end of the lecture that there is a *sine qua non* of "cure":

- That there be little (not functionally significant) or no pain
- That there be no physical restrictions and no fear of physical activity
- That all physical treatments must be discontinued

With respect to physical activity, while some of our patients never cease being physically active, the majority are intimidated by their pain and are fearful of most physical activities. This fear must be overcome, and I tell them that for years we have been advising our patients to engage in unrestricted physical activities when the pain is gone and that there has never been a report of physical trouble as a consequence. Many patients report prompt disappearance of pain, but a lot of them admit that it was months before they were able to screw up the courage to return to their previous level of physical activity.

At the end of the lecture, I give the patients a study program and instruct them to work with it for a few weeks and then call and report progress or the lack of it. At that later time, we can make the decision as to whether further work in the program is required.

Following is a reprint of the program currently in use.

DAILY STUDY PROGRAM

You have had a consultation with me. You have heard the basic lecture, and then you will probably say to yourself, "Now what do I do?"

The principle behind getting better is to understand what tension myositis syndrome (TMS) is all about and to acknowledge that TMS, and nothing else, is the cause of your pain or numbness, tingling or weakness, regardless of where you are feeling these things. You came with other diagnoses, but if you have been accepted into the program, it means that those diagnoses were not correct. You may have x-ray or MRI abnormalities but they are not the cause of your symptoms. In order to get better it is essential that you put your past medical experiences behind you and concentrate on this program. You will speed up the process if you forget everything that you were told in the past, including the diagnosis, what you are supposed to do and not to do. Concentrate on understanding TMS. You have a normal back or neck or leg or arm! Your symptoms are all due to mild oxygen deprivation, which is harmless but can cause very severe symptoms. When and where you have the pain is not important. You are going to change your brain's program by thinking about certain things every day until the pain stops.

The educational program—the lectures and my follow-up meetings—will get the job done in roughly 80 percent of people with TMS. About 20 percent will need work with one of our psychologists to complete the process. This is no disgrace since we could all profit from learning what is going on in our unconscious minds, which is what good psychotherapy is all about.

Here is the program I want you to follow:

1. If you have not already done so, read the entire book you are using a bit at a time (*Healing Back Pain or The Mind-*

body Prescription). After that, read the psychology or treatment chapter every day. Pay close attention to what you read, especially when you see things that remind you of yourself.

2. Set aside time every day, possibly fifteen minutes in the morning and thirty minutes in the evening, to review the material I am about to suggest.

3. Unconscious painful and threatening feelings are what necessitates the pain. They are inside you; you don't feel them.

4. Make a list of all the things that may be contributing to those feelings.

5. Write an essay, the longer the better, about each item on your list. This will force you to focus in depth on the emotional things of importance in your life. There are a number of possible sources of those feelings:
   a. Anger, hurt, emotional pain, and sadness generated in childhood will stay with you all your life because there is no such thing as time in the unconscious. Feelings experienced in the unconscious at any time in a person's life, including childhood, are permanent. Physical, sexual, or emotional abuse will leave large amounts of pain and sadness. But not receiving adequate emotional support, enough warmth and love will also result in anger, sorrow, and pain, maybe never felt as a child, but always there in the unconscious. Such things as excessive discipline or unreasonable expecta-

tions will also leave emotional marks. Anything that prevents a child from being a child falls into this category and should be put on your list.

b. In most people with TMS, certain personality traits make the greatest contribution to the internal emotional pain and anger. Put these at the head of your list. If you expect a great deal of yourself, if you drive yourself to be perfect, to achieve, to succeed, if you are your own severest critic, if you are very conscientious, these are likely to make you very angry inside. Sensitivity to criticism and deep-down feelings of inferiority are common and also contribute to inner anger. In fact, feelings of inferiority may be the major reason we strive to be perfect and good.

   If you have a strong need to please people, to want them to like you, or if you tend to be very helpful to everyone and anyone, if you are the caretaker type and are always worrying about your family, friends, and relatives, these drives will also make you furious inside, because that's the way the mind works. The child in our unconscious doesn't care about anyone but itself and gets angry at the pressures to be perfect and good.

c. Just as the inner mind reacts against being perfect or good, it also resents any kind of life pressure. So you should put on your list anything in your life that represents pressure or responsibility, like your job; your spouse, if you are married; your children, if you are a parent; your parents, if they are living; and, of course, any big problems that are going on in your life.

d. A subtle but important source of inner anger in some people is the fact that they are getting old and also that they are mortal. This is more common than you would think. Consciously, we rationalize; unconsciously, we are enraged. Close personal relationships, no matter how good they are, are often the source of unconscious anger, because it's very hard to be consciously angry at a parent, a spouse, or a child.

e. Add to your list those situations in which you become consciously angry (or annoyed) but cannot express it, whatever the reason may be. That *suppressed* anger is internalized and becomes part of the reservoir of rage that brings on TMS. The angers we talked about above are *repressed*—you don't feel them, you don't know they are there.

In summary, I have mentioned many things that may cause unconscious anger. Together they may be responsible for *rage* in the unconscious. But do not be alarmed. Everybody is under pressure from themselves or from life circumstances—and everybody has some degree of rage in their unconscious. This program is designed to stop the brain from producing pain because it fears that the rage, emotional pain, or sadness will manifest itself and be felt consciously if it doesn't do something to distract you. You must sit down and think about these things every day. This is the way the ideas get from your conscious mind to your unconscious mind. That's where they have to get to in order for the brain to stop the pain process.

When the pain is gone, or almost gone, start to do physical things you have been afraid to do. This is very difficult for some people with TMS, and it may take many weeks or months before you are able to get back to full physical activity. But that must be your goal in order to convince your mind that you know you have a normal back (or neck or shoulder or wherever the pain is).

Don't give up. You have to put in time and effort to make this work.

Try hard not to pay attention to your pain. When you find yourself thinking about it, force yourself to think about the psychological things on your list.

After the lectures, patients are advised to work on the program daily for three to four weeks and then call me. There is no uniform pattern of improvement. In some, it is in fits and starts; in others, it is gradual and may extend over weeks and even months. When a patient reports little or no progress after a few weeks I suggest joining one of the weekly groups I conduct, or psychotherapy, either group or individual depending on the circumstances. Occasionally, I recommend the patient attend both my weekly meetings and psychotherapy.

GROUP MEETINGS

The form of these meetings has evolved over the years. Originally, they were augmentations of my lecture. Now they follow a more traditional group therapy pattern with patients sharing their physical and psychological experiences. This gives me the opportunity to ask questions, clarify misconceptions about TMS, and reinforce therapeutic strategies. Not infrequently, patients report progress, which is an inspiration to others in the group who are struggling. Patients clearly like talking about themselves and listening to their fellow participants. Some will recover in the course of these meetings. For those who do not, psychotherapy is prescribed.

Many patients ask what to do or say to stop the pain immediately, and it is pointed out that the thrust of the program is to bring about a change in their understanding of the entire process, making it an exercise in preventive medicine rather than a method of bringing immediate relief. Once having achieved this reorganization of thought, however, one can often abort an incipient attack of pain by rehearsing in the mind the pertinent psychological factors in one's case or even "talking to your brain." I have often done this, and the following note from a colleague, Dr. Marc Sopher, illustrates how it works:

> I thought you might enjoy a short TMS story about yours truly. I was training for the Boston Marathon this winter when I had a recurrence of my TMS pain in my calf—the same that I had when I visited you last spring. I couldn't run, but I could do anything else—pretty absurd. It did coincide with an increase in stress, but I feared I was losing the battle. Running Boston was a goal, also associated with self-imposed pressure. So, despite not running for the month prior, I decided to run it. I had kept fit by riding a stationary bike and doing NordicTrack for 1.5 hours/day. I was determined to triumph over the TMS and my brain. All was well until mile 6 when the pain hit—I began an internal dialogue and told my brain that my legs could fatigue and cramp (this was a marathon, after all), but this TMS pain was ridiculous and had to leave. Within 200 yards it was gone and hasn't been back. Unfortunately, it is tough to run a marathon without running adequate mileage prior, but I finished—very tired and sore, but happy.

WHY IS KNOWLEDGE CURATIVE?

The fact that knowledge "cures" psychosomatic disorders is beyond question. But how this comes about is not entirely clear, even after all these years. Knowledge does not eliminate the rage nor change the repressed feelings that are responsible for the rage. It is not certain that even years of analytically oriented psychotherapy could bring about those changes. But thousands have become pain free simply by reading my books, and thousands more have gone through my program successfully, most without the need for psychotherapy.

It has been postulated that the purpose of symptoms, whether of TMS or any of its equivalents, is to prevent the unconscious experience of rage and painful emotions from becoming conscious, overt. Logically, one would conclude that the only way to stop painful symptoms would be to eliminate the threatening unconscious emotions or get them to explode into consciousness. But neither of these can be done, and yet the education process can stop the pain. There is only one possible explanation. We know from experience that the theoretical wall, the barrier separating the conscious from the unconscious mind, cannot be breached from below—that is, the rage will not break through into consciousness—but there is nothing to stop us from intellectually breaching the barrier from above, from saying, "I can use my imagination and think of my unconscious as the basement of my mind. I know what's down there even though I can't see or hear it. I have been taught to recognize the inhabitants of my unconscious mind, in all of their dangerous, unflattering detail."

And the unconscious response might be, "He/she has found out what he/she wasn't supposed to know, and has discovered the closely guarded secret about the rage and other painful or threatening emotions, like hurt and sadness, that go back to childhood. The cover has been blown on this covert operation, so there's no sense in continuing the pain."

It is obvious that the educational process effects a "cure." For patients with mild symptoms, the mere knowledge that the symptoms are psychosomatic, without further psychological insight, is curative in and of itself and undoubtedly explains many of the book "cures." The additional pieces of information about the nature of the pathophysiologic process, the workings of the unconscious, the existence of internal feelings, and the consideration of factors contributing to those feelings, most particularly one's own personality characteristics, all have great therapeutic power.

Here is a quote from a book on Alfred Adler's work by Heinz L. Ansbacher and Rowena R. Ansbacher that indicates that Adler had similar ideas on the role of therapy:

> Thus successful therapy is not the result of better repression by a so-called strengthened ego. Rather, it is the outcome of an insightful reorganization of their thinking, which must then lead to a reorganization of what is going on in the unconscious to the extent that the psyche is no longer threatened by the rage to the same degree it was prior to the teaching. The magnitude of the threat characterized the old organization; the educational process has diminished the threat.

This is another explanation of what we see in practice. I am open to any other interpretation of the facts.

A very bright young wife of a patient suggested that causing TMS was a "trick" of the brain and, having learned about the trick, it no longer works. Very ingenious.

Knowledge, awareness, and insight have been the cornerstones of analytically oriented psychotherapy since Freud, so it should come as no surprise that they are the keys to treating psychosomatic disorders.

How does one explain the fact that some people are "cured" simply by reading one of my books, that some need to see the doctor and go through a formal program, and that others additionally need psychotherapy? Perhaps the best explanation is that there are varying intensities of the unconscious states that demand and cause psychosomatic symptoms. Another factor may be the depth and power of the repression. In the mildest cases, merely acknowledging that the pain is psychologically rather than "physically" induced is enough to reverse the process. With increasing severity of pain more and more intervention is required.

PSYCHOTHERAPY

Approximately 20 percent of those accepted into the program need psychotherapy to get better. In an ideal society, psychotherapy would be an integral part of our program and be covered by health insurance. In our socially antediluvian culture it is poorly covered by insurance, if at all, and represents a great financial burden for many patients. Indeed, learning about the unconscious mind should be part of our educational systems, for it is fully as important as reading, writing, and arithmetic.

Failure to improve is the indication for therapy. Such therapy implies that the reasons for the internal feelings are strong and will not respond to simple recognition. Many patients will deny the existence of rage, for example, at a parent. Others are unable to feel sadness, anger, disappointment, abandonment, and so on. It is clear that the therapist must be dynamically (analytically) trained to recognize and work with these problems, and the process is often prolonged.

It was pointed out in chapter 3 that our program has seen many people who would probably never have come to psychotherapy but for the recognition that their pain was psychosomatic in origin. This is a testament to the importance of recognizing these disorders and to the tragedy that follows the failure to do so.

Arlene Feinblatt, Ph.D., has been my colleague and coworker for over thirty years. By dint of circumstance, she is a pioneer in the development of psychotherapy for psychosomatic disorders. Since no one in the field of psychology or psychiatry has had extensive experience with the musculoskeletal pain of psychosomatic origin, and since as a consequence there is no guiding literature on the subject, and in view of the fact that this ailment represents a public health problem of major proportions, it was essential to develop appropriate psychotherapy for these patients. Dr. Feinblatt has done that job admirably. She has also trained a large cadre of therapists over the years.

I have said elsewhere that it is essential for patient acceptance that a psychosomatic diagnosis be made by a physician. It is equally important that the physician and psychotherapist remain in communication throughout the course of treatment. The responsibilities are clearly delineated: the physician's role is to educate patients on the nature of the psychosomatic process, essential for successful treatment whether or not psychotherapy is prescribed, and precedes the referral. The physician manages physical problems and answers questions that may arise during the course of psychotherapy. The psychologist, of course, has the major responsibility of diagnosing and treating the psychodynamics responsible for symptoms. The psychologist also determines whether psychoactive medication is necessary as an adjunct to therapy and refers the patient to a psychopharmacologist should that be the case.

The following section describing the psychotherapeutic program was prepared by Dr. Feinblatt.

SHORT-TERM DYNAMIC PSYCHOTHERAPY

Since psychosomatic disorders are the result of unconscious processes and conflicts, our psychological program has focused on the use of short-term dynamic psychotherapy as the best means to deal with underlying stress and emotional conflict. Our psychological program involves examining the effects and interactions of psychological factors on the body and then connecting patients' emotions with their physical reactions. The major emphasis in our treatment is to uncover defenses and the repressing affects. Since behavioral inhibition, repression, and the effects of disclosure have been found to be interconnected with physical processes, this method of treatment appears particularly well suited as treatment for these patients.

CONSULTATION/EVALUATION

Psychotherapy for patients with TMS begins with the psychological consultation/evaluation of each patient. Following the physician's referral, each patient is seen for an initial session, which may consist of one to two forty-five minute interviews. Background health, family, education, vocation, and social history are explored. This examination of historical factors, important as they may be, is meant to be only the starting point for the psychologist's exploration of the patient's ego strength, reality testing, cognitive ability, defense mechanisms, nonverbal behavior, and ability to maintain a rapport with the therapist.

The process of the consultation/evaluation interview is determined, in large part, by the patient's response. The pressure from the interviewing psychologist and the degree of toleration of the patient to the stress of the process serve to identify the patient's strengths and weaknesses. These, in turn, determine what kind of therapy—none, individual, or short-term group—is suitable for each patient.

The consultation is basically a microanalysis, and results will determine treatment recommendations. As such, it requires great skill and experience to utilize this method of examination.

Observation of patients begins even before their consultation takes place. We note patients' general demeanor in preliminary telephone contacts, attitudes toward the waiting room atmosphere, and whatever other contacts we have with them. We look for indications of eagerness, anxiety, suspiciousness, or compulsivity.

We pay close attention to nonverbal behavior during the interview. Patients who fidget or are unable to tolerate sitting during the entire interview, as well as the manner in which they deal with this inability, provide the interviewer with significant information about the patient's ability to relate and general personality style. Since patients with psychosomatic disorders exhibit repressive personality styles, they often have difficulty maintaining eye contact with the interviewer and/or often laugh or smile when relating painful or difficult material. The interviewer challenges these repressive behaviors in order to gauge the degree to which these defenses are entrenched. The degree to which the patient tolerates these challenges and the patient's responses to them will indicate the ego strength, and ability of the patient to make use of further therapy. For those patients unable to tolerate direct challenges, a more supportive, less anxiety-provoking, and more prolonged treatment may be necessary.

Following the initial evaluation, the psychologist decides whether treatment within our program is appropriate. If it is found not to be appropriate, the psychologist may refer the patient to either psychiatric treatment or supportive psychotherapy, or both. When our program is deemed appropriate for the patient, the psychologist will prescribe either short-term group or individual therapy, depending on the evaluation of the patient.

SHORT-TERM GROUP PSYCHOTHERAPY

A recommendation of short-term group psychotherapy is made when patients appear to have little idea of the degree to which they utilize repression or if they are unsure of some of the psychological factors affecting them. Short-term group may also be suggested for patients who have never been in therapy and are unaware of the psychotherapeutic process. Group therapy is also recommended for those patients who have difficulty relating the mindbody connection to themselves. Each short-term pain group consists of eight weekly ninety-minute sessions, led by the senior therapist and, usually, a cotherapist and somewhere between five and ten patients. If individual psychotherapy is recommended, the choice of therapist is determined by a variety of factors that may include years of experience, cultural background, and personality style.

The first group sessions are devoted to identifying, clarifying, and challenging the defensive maneuvers patients employ. Most patients are unaware of these defenses and are unable to even define their feelings as distinguished from their thoughts. In addition, they often confuse one emotion with another. Such patients are often labeled "alexithymic," implying that they are not capable of generating feelings, a concept with which we strongly disagree.

In the middle weeks of treatment the focus shifts to emotional repression and the mechanisms by which this leads to physical symptoms. Patients begin to connect a variety of emotional states with their physical concomitants. Sometimes patients who are unaware of the physical responses of their bodies may be confused between somatic (physical) defenses, anxiety, and their normal impulses and feelings. It is during these weeks that group members usually begin to exhibit greater cohesion and begin to learn from each other's experiences.

The major issues taken up during the final sessions are loss and grief. These evolve naturally as the group experience matures.

The efficacy of short-term dynamic group therapy has been documented by two large studies, both of which found significant reduction in intensity, duration, and frequency of pain, as well as reduced sleep disturbances in participating patients.

INDIVIDUAL THERAPY

All our psychotherapy is administered within the same framework. Whether a patient is in group or individual psychotherapy, the therapist's job is the same: to help the patients attain a greater understanding of the defensive structure that is shielding their conscious selves from the destructive aspects of their unconscious rage, and to increase their emotional awareness overall. To that end, treatment is tightly focused. In short-term dynamic therapy, patients are constantly challenged to justify inappropriate behavior. For example, if a patient laughs or smiles in reaction to stimuli that clearly call for a different response, the therapist will draw the patient's attention to that fact and ask the patient to explain such a nonrational anomaly. Such emotional incongruities are important clues to clarifying the patient's understanding of his/her own emotional profile. At each of these points in treatment, therapists must intercede to increase patients' awareness between underlying emotions and their ability or failure to express them. We also confront all attempts at denial or rationalization.

At times, patients will attempt to describe their inappropriate behavior as an example of their emotional awareness, and the therapist must help the patient to understand that it is nothing of the sort. Such an interpretation is, in fact, an example of avoidance rather than an actual experience of emotional truth. When challenged by the therapist, a patient's reaction—almost always evasive or self-defeating or irrational—can be employed as a method of intensifying the emotional experience. At other times, patients will use silence as a way of decreasing their sense of vulnerability or, in their minds, expressing their rage. The therapist can help the patient explore the consequences of such behavior, and can sometimes add a level of urgency by pointing out that the continuation of such behavior will tend to increase the length of treatment.

Treatment must focus on people's psychologies rather than their physical symptoms. When patients introduce medical or physical phenomena into the discussion, the therapist must immediately respond with questions about the psychological aspects of the patient's life prior to, or at the time of, the symptom's appearance, as well as other psychological issues. For example, if a patient associates the onset of pain with a long car trip, the therapist should pose questions regarding the circumstances of the trip, the purpose of the trip, the patient's state of mind about making the trip, as well as the emotional outcome of the trip. Awareness and understanding of the emotional underpinnings of defensive structures must occur during treatment, and, as noted earlier, physical and repressive measures must be recognized as counterproductive.

Because our therapeutic methods are deliberately challenging, patients who have already had experience in psychotherapy with other mental health professionals have to adjust to the more rigorous character of our program. Almost all such patients report a sense of anxiety yet relief about the changes in treatment. They tell us that although they feel more anxious, they are desirous of moving more quickly toward symptom resolution. In fact, an increase in consciously experienced anxiety often occurs prior to or during symptom reduction. Therefore, although symptoms (such as back pain) create motivation for therapy, they can also be used as a yardstick by both the patient and the therapist to measure their progress or lack of it in treatment.

Patients must view treatment as a partnership in which they are regarded as fully empowered adults. A frequent occurrence during the process of successful psychotherapy is a fluctuation of symptoms, such as a reduction in pain followed by an increase in pain soon after. This is a more common pattern than a sudden or gradual cessation of symptoms. These changes may reflect the cycle of catharsis and resistance as well as the fluctuations in the patients' understanding of the true nature of what is causing their disorder, as their old incorrect or self-defeating rationales give way to more adaptive defenses. It can take time to fully comprehend the complicated interaction in which your buried rage brings on feelings of guilt which in turn can engender low self-esteem.

Some patients report that their symptoms (usually pain) have begun to move from one place to another. This may be taken as a favorable sign that therapy is having a beneficial effect and can also be used to reassure patients that such a pattern is more consistent with psychological rather than disease-induced symptoms.

One of the greatest obstacles therapists face in the treatment of the psychosomatic patient is their patients' mistaken assumption that they know what the underlying problems are and what has caused their disorder.

Patients often come to treatment experiencing guilt regarding their dependency; anger at the lack of understanding by others of their symptoms, which are frequently invisible; and frustrated by the lack of progress they have made with conventional medical treatments. These feelings are more easily accessible and can often be utilized to begin the work of developing greater emotional awareness, which will in turn lead to greater access to the unconscious.

For certain patients, therapy must be modified to reduce the degree of anxiety engendered by the therapist's relentless challenge. These include patients who are severely or suicidally depressed, patients with brain damage, and patients who see themselves as getting worse rather than better. For these patients, treatment must be more supportive in nature so that their ability to attend is not overwhelmed by anxiety.

Vigilance is the therapist's most effective tool for treatment. All nonverbal as well as verbal cues must be utilized. Postural muscles, facial muscles, and breathing are monitored. When patients' underlying feelings are recognized by the therapist, patients frequently express relief and become aware of a reduction of physical tension. This reduction marks a significant moment in treatment. The patient is now able to distinguish between physical and psychological defenses. It also serves to illustrate to the patient how the therapist's relentless challenge is actually helping the patient.

In general, the psychosomatic patients with whom we have worked suffer from great anxiety and fear. They often feel victimized, and some have, in fact, often been victims of sexual or physical abuse. Often, as a result of their histories, many patients are acutely sensitive to the needs of others. They suffer from low self-regard and are driven to succeed and take care of others. They have a need to feel in control of situations about which they ultimately feel totally out of control. This often occurs as a result of feeling that their own bodies have betrayed them. They frequently have a history of other psychosomatic disorders. Some have lost a parent or have experienced the domination of a demanding parent. Patients with psychosomatic disorders frequently display other self-punitive behaviors, usually in the form of being unduly demanding on themselves. One such example was a patient who engaged in excessive exercise, spending at least six hours a day on the track or the treadmills or the weight room, with the sole goal of exhausting himself.

Psychosomatic individuals often have a troubled history with other medical health professionals. They may have been misunderstood due to the fact that their problems are frequently invisible.

They have met with significant frustration for themselves as well as the individuals who have tried unsuccessfully to treat them. As often as not they have dropped out, removed themselves from normal social contacts, and/or become more dependent on others, usually family members. The result has been guilt and lowered self-esteem, which in turn serves to increase their psychosomatic symptoms. When patients are finally able to lose their symptoms, they achieve greater self-understanding and deal more effectively with others. One of the most satisfying aspects of a psychotherapist's work is that successful treatment of the psychosomatic patient results in a healthier emotional as well as healthier physical individual.

One of Dr. Feinblatt's early trainees was Dr. Eric Sherman. The following cases are from his practice.

ABNER'S CASE

When Abner, a thirty-five-year-old man, came in with a history of severe low back and leg pain, he had already seen several orthopedists who recommended surgery for a herniated disk that had shown up on an MRI scan. He was frightened by the prospect of surgery, as well as by the disappointing results of friends and relatives who had undergone similar procedures. Abner clearly appeared anguished, and his physical activity was significantly limited. One member of Abner's church was alarmed by the contrast between his current physical state and what had been his hard-driving and indefatigable style. On learning of the back pain and the recommendation for surgery, the acquaintance confided in Abner about his own similar struggle with incapacitating back pain. He described how symptoms began to improve after consulting with Dr. Sarno and participating in his therapeutic program. At this point, Abner was desperate and willing to consider anything, so he pursued his friend's recommendation to see Dr. Sarno. Being an extremely intelligent person and able to master complex and novel information, Abner quickly grasped Dr. Sarno's central premise that unconscious anger plays a critical role in the development of back pain.

The unique problem for Abner was that he didn't consider himself emotionally inhibited. In fact, Abner and those closest to him would agree that he was angry most of the time. They knew him to be a hotheaded, argumentative man, who not only sought out and provoked confrontations, but seemingly relished them regardless of the outcome. Therefore, Abner feared that Dr. Sarno's diagnosis would not apply to his case. He was unaware of the distinction between conscious and unconscious anger. He began to resign himself to surgery. However, the very fear of surgery helped him overcome his misgivings about the accuracy of Dr. Sarno's diagnosis. He accepted that his pain symptomatology was a psychosomatic disorder and reluctantly followed Dr. Sarno's recommendation for individual psychotherapy.

Abner, a highly engaging and dauntingly articulate man, began treatment boasting in meticulous detail about his most recent angry outbursts. He was genuinely baffled how someone like him could possibly be denying angry feelings. He therefore questioned how he could possibly be helped by treatment. As the therapy proceeded, it became apparent that Abner overreacted to ordinary indignities we all endure in the course of our everyday lives. Though still unconvinced that treatment could help him, Abner was nevertheless able to develop insight into his disproportionately intense reactions to what even he agreed were generally trivial provocations. He became curious about why he felt impelled to "kill houseflies with howitzers," even when the potential benefits could hardly justify the amount of time, effort, and energy he invested in these battles.

Abner began to observe that whenever he perceived slights, he experienced the other person as "dissing" him, essentially conveying to Abner that he was unimportant and unworthy of consideration. On his own, Abner realized quickly that these feelings were painfully familiar to him. His own mother was so preoccupied with her mother's ill health and needs for caretaking that she often neglected Abner's emotional needs, frequently delegating his care to others. When Abner, craving his mother's attention, would naturally protest, he was scolded for being selfish, thinking about himself instead of his sick grandmother.

To make matters worse, Abner's father, a businessman, was extremely grandiose. He would buttress his precarious sense of self by subjecting the family to gassy accounts of his derring-do in the business world. Whenever Abner would seek his father's attention on these occasions, he would be upbraided for not listening. If Abner failed to support his parents' narcissism, he would be punished for being selfish and self-centered. The cruel irony of this warped family dynamic infuriated Abner. He railed against the injustice of a child's having to sacrifice himself in order to protect his parents. His understanding of this dilemma deepened as Abner realized that this self-sacrifice had been essential for his own survival. If he hadn't followed the rules, he would have been less likely to receive even the crumbs of parental attention he barely subsisted on.

With this realization Abner revised his sense of himself in a significant manner. Although he was often combustible toward others, he now appreciated how extremely inhibited he was when it came to experiencing angry feelings toward his parents. He feared he would inevitably act on his rage even when he experienced it solely on a private, internal, emotional basis. Therefore, his feelings could threaten his very survival by pushing him to "bite the hand that feeds him."

How could he possibly love his parents if he felt such anger toward them? In Abner's mind, only a monster would not love his parents. For him, there was no such thing as ambivalence; you either loved someone or hated him. Therefore, Abner avoided creating situations that would reactivate his unconscious anger toward his parents. Just as when he was a child, the adult Abner deftly side-stepped the need for his parents' attention. In this way he could prevent the eruption of his unconscious rage, the very rage that was originally ignited by being deprived of his parents' emotional involvement. However, the avoidance of conflict was at best a temporizing solution, because it couldn't begin to contain his all-consuming rage. So he redoubled his efforts to muzzle the expression of his anger by developing excruciating pain. Severe pain served as a powerful distraction from intolerable and terrifying feelings of destructive wrath. The pain protected Abner's sense of himself as a loving son without limiting his physical activity—to wit, his ability to function sexually. Obviously, physical suffering can only be a tenable solution when it's the single alternative to unleashing an anger destructive enough to jeopardize one's very survival.

The treatment for psychosomatic pain syndromes does not require that patients renounce particular feelings. As Abner learned during the course of treatment, feelings are unbidden experiences. We have no control over what we feel, but we can and must exercise control over how we respond to our feelings. The goal of treatment, then, is to enable the patient to respond to his emotional conflicts more adaptively—by means other than developing pain.

At the outset of treatment, Abner could not recognize his own angry feelings toward his parents because he was not comfortable with such emotional experiences on a private and internal basis. As Abner understood that angry feelings toward loved ones do not represent an absence of love, he was better able to tolerate the feelings and reflect on them. Once he was able to maintain his self-esteem while experiencing anger, his pain decreased significantly. Abner eventually became completely asymptomatic with respect to his pain for a period of five years.

Then, in the midst of a family crisis, Abner's pain recurred. He was caught up in a drama replete with betrayals, recriminations, and anguish usually encountered only in Greek tragedies or soap operas on television. His pain was so severe that he required staggering doses of painkillers for relief. And that relief was barely adequate and often short lived. Abner's family insisted he consult the orthopedists who had recommended surgery almost six years before. His family concluded some new physical disorder must have developed after so long a time without back pain. Abner bowed to their pressures. Privately, he too was fearful that this episode was different. Finally, however, Abner concluded that his symptoms could only represent a new outbreak of psychosomatic back pain despite what doctors and family were telling him. His realization came from the recognition that he was entirely free of pain during sexual activities (which for him often verged on the acrobatic). A repeat MRI also demonstrated no interval change from the original done six years earlier.

In this latest incident, Abner was indeed painfully aware of how angry he felt toward his mother and father for their reckless behavior, which had plunged the entire family into a crisis. Nevertheless, he could not discuss these feelings with his father, whom he held mostly responsible for the calamity. Abner feared if he revealed his feelings, his father would be demolished. He observed that the loss of his father would terrify him at least as much as it would sadden him. Abner's fears of dependency finally emerged.

The severity of Abner's pain symptomatology had alarmed everyone who knew him. He appeared to be a broken man, physically, emotionally, and spiritually. Abner soon came to realize how frightened he was by the magnitude of his rage. He rendered himself harmless by incapacitating himself with excruciating pain. That way he could no longer threaten his father with lethal rage, jeopardizing his own survival. Instead, he could indict his father with physical suffering, disarming the father at the same time. No matter how angry his father became in response to his son's veiled accusations, the father could not retaliate with the threat of abandonment when his son was in such dire straits.

As the body language of physical suffering was translated into its emotional substrates, Abner was able to reflect on his fantasies of destructive rage and emotional devastation. Once again, Abner recognized how he protected his father's grandiosity by perpetuating his own sense of dependence. Despite Abner's prodigious business acumen, he disparaged himself as an impostor—"a boy sent to do a man's job." Abner ensured his own survival by not "raining on his father's parade." Unfortunately, Abner's legitimate wish to develop his abilities became confused with his fears of hurting his father.

As Abner increased his tolerance for his angry and dependent feelings, his pain subsided, although not as dramatically as during the first course of treatment. Eventually, Abner was minimally symptomatic on a consistent basis, and he decided to terminate treatment, knowing that more work awaited him.

HIROKU'S CASE

Hiroku is a twenty-five-year-old Japanese American insurance industry executive with a long history of unexplained, intermittent medical conditions. These conditions have included pains in her shoulder, elbow pain, headaches, and assorted symptoms of gastrointestinal distress. Most recently, she had developed right foot pain, unrelated to any physical trauma. All of these conditions had been dealt with extensively by the most renowned practitioners in their fields, no doubt owing to the fact that both the patient's mother and father are physicians. The patient's mother has always assumed her daughter's complaints were psychosomatic disorders, unless proven otherwise. Indeed, although the physicians' evaluation of Hiroku never diagnosed her with a psychosomatic condition, they simply demurred and reported that they found nothing to explain her problems.

Hiroku's mother attended a professional conference on mind-body issues in oncology. One of the presenters cited Dr. Sarno's contributions to the field of psychosomatic medicine. Hiroku's mother urged her daughter to consult with Dr. Sarno, which she did.

In her meeting with Dr. Sarno, Hiroku showed herself to be extremely perceptive and articulate. She cataloged her emotional conflicts in a breezy, engaging, and, most of all, sophisticated manner. Although she agreed with Dr. Sarno's premises, and frequently recognized herself in the pages of his books, she was somewhat skeptical as to whether the treatment would work for her since she was already so in touch with her feelings. As an example, she launched into a devastating tirade against her boyfriend without even breaking a sweat. Her aplomb was impressive, but more eerie than admirable. Despite her skepticism, she was highly receptive to Dr. Sarno's recommendation to begin psychotherapy.

Within the first few sessions Hiroku's foot pain almost completely resolved, for reasons that still remain unfathomable. Then the honeymoon was over—abruptly. Hiroku's foot pain not only worsened, she developed new symptoms in her knee and wrist, along with a rash on her forehead. Hiroku demanded answers and resurrected her early skepticism about the treatment's efficacy.

When questioned about recent experiences, Hiroku freely reported feeling angry about out-of-town visitors and how the requirements of hospitality further strained her already overscheduled life. In fact, this very access to her angry feelings proved that her treatment was doomed. It was then explained to Hiroku that while many people are fully aware of experiencing emotions, a part of them struggles against feelings considered taboo. For example, many mothers are aware of resenting the demands of newborn babies, yet they often feel guilty. They are ashamed of having these feelings, even when their behavior toward the child remains beyond reproach. Hiroku then acknowledged that although she outwardly appeared gracious toward her guests during their entire stay, she felt ashamed of her resentment toward them. According to Hiroku, to experience resentment was tantamount to being selfish, no matter how you actually behaved.

Hiroku continued to deepen her understanding of how she couldn't tolerate certain emotional experiences, even on a private and internal level. She began to use her foot pain as a guide to introspection, rather than a clue to unraveling a physical medical mystery. Whenever she experienced foot pain, she would first ask herself, "What am I feeling?" However, she soon encountered an obstacle—the answers to her question were often unacceptable to her. For example, when she identified her anger toward a supervisor who unfairly criticized her, she experienced self-contempt for the very weakness, which made her vulnerable to criticism in the first place.

Overall, Hiroku's pain symptomatology remains significantly improved. Since Hiroku is in the earliest stages of treatment, further progress will no doubt be related to a deeper understanding of the interplay between her feelings of anger and neediness.

LIAM'S CASE

Liam was almost thirty-four years old when he initially arrived for treatment of severe, psychosomatic back pain. Liam is the oldest of four boys. He is also the first born and a much adored "Christ-grandchild" in a large European family. Self-described as an army brat, Liam has lived all over the world and is dazzlingly fluent in three languages. He oozes the easy self-assurance of someone who has "been there, done that." A high achiever and well liked by everyone, Liam never has an unkind word for anyone, at least not out loud.

Liam's first episode of disabling back pain occurred on a return flight from a vacation. He unexpectedly ran into his former fiancée's best girlfriend. The woman updated Liam with news that his former fiancée was engaged to another man and would be marrying later that year. Although nothing was explicitly stated, Liam "just knew" that this woman noticed that he was unattached and "couldn't wait" to now go and tell his former fiancée the good news about the pathetic state of Liam's love life. When Liam helped the woman stow her baggage in an overhead bin, he was stricken with severe back spasms that caused him to fall to the floor in the aisle of the airplane. At the time, Liam concluded that he must have "pulled something." Braced by several stiff drinks, he endured the three-and-a-half-hour flight, hoping that his back pain was just some freak occurrence. After all, he was an extremely healthy thirty-year-old and very athletic. How could anything as innocuous as stowing a small suitcase result in such a pain?

After several agonizing days at home, Liam consulted a physician, who referred him to an orthopedist. The orthopedist ordered an MRI scan, which revealed a herniated disc, and the orthopedist subsequently advised a conservative treatment with bed rest, anti-inflammatory drugs, and possible physical therapy, depending on Liam's response to the other recommendations. Liam's condition only marginally improved, and he was referred to a physical therapist for more aggressive treatment. He gradually improved and eventually discontinued the physical therapy on his own. He remained entirely free of back pain for almost four years, at which point he suffered a recurrence.

In this instance, the orthopedists he consulted all recommended surgery. When he learned that his latest MRI was indistinguishable from the original, he questioned the need for surgery. After all, he said, if nothing had changed in his spine and he had remained asymptomatic for four years without surgery, why wasn't it reasonable to assume the same thing had happened again? To a man, the doctors dismissed his question and invoked his previous history of back pain as the basis for recommending surgery.

Liam systematically availed himself of almost every known alternative treatment and experienced varying degrees of improvement. While searching a bookstore for yet more information, he happened upon several of Dr. Sarno's books. As he read them, he recognized himself on almost every page. He also realized in hindsight that his first episode of back pain was almost certainly precipitated by an avalanche of unresolved feelings surrounding his broken engagement.

Liam entered psychotherapy. It wasn't until the better part of a year had elapsed that he enjoyed any improvement in his pain symptomatology. At that point he was able to formulate an understanding of why it had been imperative for him that certain emotional conflicts be banished from his conscious awareness. Although Liam was consciously aware of intensely angry feelings toward his father, he was deeply ashamed of having these feelings. Adding to his shame was his belief that he was the only one in his family who experienced such feelings of hatred. On the rare occasions when Liam voiced protest or disagreement with his father, his father would dramatically flaunt his outrage and emotional injury, verbally bludgeoning Liam into recanting his blasphemous sentiments. Of course, he erroneously interpreted his mother's passivity and his brothers' paralysis as agreement with his father's position.

Like many patients suffering from psychosomatic pain syndromes, Liam is a high achiever and much admired. He is the one member of the family that everyone looks up to. He is the consummate caretaker. He has been filling this role since his earliest childhood. He translated for his parents and explained American culture to them. He interceded for his brothers when cultural clashes inevitably occurred. And he consoled his mother for her marital disappointments by being the best son any mother could ever ask for. Liam's self-esteem was entirely predicated on successfully taking care of others, especially at his own expense.

As treatment progressed, it became clearer that if Liam were to believe that other members of his family shared his feelings toward his father, then he could see that his anger represented a legitimate reaction to his father's tyranny, and not a sign of a son's disloyalty and ingratitude. The recognition of the father's abuse, however, was followed by Liam's self-loathing for having failed to protect his mother and brothers from the same tyranny. For a long time, it had been more comfortable for Liam to feel shame for his angry feelings than shame for his scared and helpless feelings.

Liam has remained completely free of TMS for more than three years, despite having experienced a series of tragic events during that time. According to Liam, "I can now feel shitty, without feeling like I'm shit." And because Liam can merely feel "shitty," he doesn't experience back pain.

Following is the experience of another psychotherapist, Robert Paul Evans, Ph.D.

With all the wonderful new scientific discoveries, we are in the midst of a shift to a new paradigm, an expansion in the way we perceive and experience ourselves and the universe. However, the idea that there is a mindbody connection is still a foreign and at times scary concept for most people, including many practicing professionals. In fact, to talk of a mind-body connection is misleading, for it suggests two separate distinct units somehow "connected" rather than one integrated whole being. As Dr. Sarno has done, I shall refer to this whole entity as mindbody (unhyphenated).

Webster's Dictionary defines healing as to "make whole." For many patients with TMS in particular, the very first obstacle in the healing process is their disbelief, or at best, their ambivalence with regard to this integrative, holistic view of mindbody. It is, therefore, important to address this issue on an experiential level at the very outset of treatment. People with TMS need to be able to experience themselves in the act of repression of an emotion. This experience allows them to witness firsthand the mindbody at work, such that they can begin to embrace difficult emotions and facilitate healing. What follows is a brief description of my initial psychotherapeutic approach to working with TMS—what I do to help facilitate this healing, how I do it, and how at times it differs from more traditional forms of psychotherapy.

The thought that came to me first was a quote from Victor Hugo, "Lose your mind and come to your senses." In my view, this thought approaches the essence of psychological treatment for people with TMS. Victor Hugo had it partially right. We don't want to lose our minds but rather learn how to experience the mind and body as they truly are—integrated as one. Coming to one's senses means letting go of the tight grip the unconscious mindbody has such that it results in physical symptoms like TMS or reflux. The purpose of these symptoms is not to hurt but to distract us and thus to protect us from experiencing what is unconsciously perceived to be more painful or unpleasant or what we have come to believe are unacceptable emotions. When anxiety, fear, anger, rage, shame, guilt, hurt, sadness, sorrow, perceived undeserved joy, and the like are overwhelming in intensity, they penetrate or bypass the common psychological defense mechanisms. We experience physical symptoms instead of these emotions because the unconscious mindbody considers them to be less painful, less dangerous or harmful than the emotions. That may be hard to believe, especially when the pain is severe, but the severity demonstrates the intensity and power of those repressed feelings and the fear that they engender in the unconscious. When a patient can observe him/herself in the act of repressing a given emotion in a given moment and almost simultaneously begin to experience the emotion itself, it allows him/her to feel more confident in the mindbody experience, which in turn facilitates healing. One problem, however, is that one cannot experience something that, by definition, one is not aware of. Therefore, techniques like the following are designed to facilitate this process. The observing part of the individual can learn to become, as I like to say, "more comfortable with the discomfort" of what the participating part of the person is experiencing emotionally. Gradually, the embraced emotion is experienced not just on an intellectual level but in a full holistic, integrated way. In the early stages of therapy it is common to see that when some distressing emotion begins to surface and threatens to be felt consciously, the unconscious mindbody may make the pain worse or give the person another symptom to pay attention to. The therapist's task is to help patients develop and utilize techniques that will bring them "to their senses," that is, to experience mindbody integrated as one. When that has been achieved, TMS symptoms are no longer necessary. Before we can deal with traditional types of psychodynamic issues, we must first address the mindbody "split." Once the TMS patient feels more comfortable with the experience of this holistic integration and can begin to embrace once intolerable emotions, we can begin to do other, more traditional psychotherapeutic work that may be necessary, such as anger or post-trauma management, or dealing with more subtle issues such as low self-esteem. All of these play a part in the original personal mindbody disconnect, which is later reinforced by the culture we live in.

Though the goal is clear, the therapist must be careful in the early days of treatment to have respect for the physical symptoms, to find a balance between gentle confrontation of the defenses that protect us from feeling the uncomfortable, painful emotions, on the one hand, and creating and maintaining an environment in which we can feel safe and secure enough to begin to experience those feelings. Although the need for a safe environment is important in all forms of psychotherapy, it is especially crucial in working with people with TMS in view of the immediacy of the clinical situation. The therapist must keep in mind that people have been in pain, often for a long time, and are looking for fast relief. Their defenses have become well established and must be dealt with immediately. So they must develop trust in the therapist, feel safe and secure, and, therefore, be able to tolerate the gentle interruptions that are introduced in order to begin to penetrate those defenses.

Following are some case histories that illustrate typical defenses and how they are approached therapeutically. Names and other nonessential factors have been changed.

Susan was a forty-six-year-old married woman who demonstrated how unconsciously exhibited facial expressions or verbalization can serve as defenses. During her first session she began to talk of how her father used to yell and shout at her, frightening her, and eliciting from her an apologetic rather than an angry response. At the precise moment she mentioned feeling annoyed at her father she chuckled ever so slightly. When I gently interrupted and asked her if she had noticed the chuckle, she said she had not. Susan chuckled because at that exact instant she was beginning to sense the emotion of anger on a mindbody level rising up from the unconscious in response to her conscious acknowledgment. As this emotion began to surface, she became nervous and uncomfortable. The chuckle was a defense protecting her from experiencing further discomfort. By chuckling inadvertently at that moment, Susan automatically repressed that undesirable emotion so as not to feel any further discomfort attached to it. The chuckle was a clue to what was going on unconsciously. It was therapeutically essential to interrupt and call her attention to it so she could begin to allow herself to simultaneously observe and feel what she had repressed all her life. The goal of embracing the once repressed emotion is not simply for the sake of experiencing it, but to learn that over time the "adult observer" can tolerate and ultimately integrate the emotion in a new, healthier way. For Susan, the anger became increasingly less scary and unbearable and came to be recognized as normal and natural. She was learning how to reconnect and integrate mind and body. As she continued the process, the TMS pain receded.

One deviation from a more traditional therapeutic approach was the interruption of her story so that she could have a comprehensive emotional-intellectual (mindbody) experience rather than an intellectual one only. If Susan had continued to elaborate on the reasons for anger at her father, we would have had additional information, but the more important goal of embracing the emotion would have been lost. My calling her attention to the incongruous behavior, the chuckle, caused her to observe the repression while feeling the anger, the kind of experience that patients say brings a sense of relief. Over time and with practice, Susan's TMS pain subsided completely.

Another kind of defense is the use of what I call buffer words or phrases, such as *probably*, *maybe*, *sort of*, and the like. Sometimes inadvertently switching from first person to third person works as a buffer. Any word, phrase, or grammatical usage that succeeds in protecting the person from experiencing an uncomfortable or painful unconscious emotion is a buffer. As with the chuckle, these defenses have to be immediately recognized and gently brought to the person's attention in order to enhance the ability to observe the mindbody disconnect and thus embrace a repressed feeling.

My therapeutic experience with Alice is a good example of the use of buffers. She was married with children, bright, insightful, and possessed of a good intellectual understanding of some of her problems. During a session she began to talk about her father, children, and siblings. At one point she said, "I think I'm probably angry at my husband." *Probably* is a buffer word. Rather than allowing her to continue, I interrupted gently and repeated the word *probably* as a question. Put on the defensive, Alice immediately revised her comment and said, "I guess I'm very angry at him." The use of the word *very* may have been an indication that Alice was beginning to recognize the intensity of her anger but was still buffering the emotion when she said, "I guess." Conceivably, Alice's unconscious mindbody hoped that I would stop questioning her if she admitted to being very angry.

Anger is often frightening for people with TMS to embrace because of either the fear of losing control (the perfectionist) or significant worry and concern that the object of the anger will dislike or reject them (the goodist). At other times, embracing anger can be overwhelming because of the possibility of profound hurt and sorrow that often underlies the anger. In these cases anger is like TMS symptoms in that it also protects the person from even more painful emotions.

It was important to continue gently penetrating Alice's defense against embracing the anger, so I ignored the word *very* and questioned the phrase *I guess*. A little exasperated, Alice then said, "I'm angry at my husband." No buffers now. She had taken an important first step in observing herself experience the use of buffers to help her disconnect. This observation led her to embrace the anger in a mindbody way. She expressed some fear at what she had done but also experienced a sense of relief and a release of some tension and pain. This was a beginning; she knew there was more work to do, but the release of tension and temporary reduction in pain told us we were on the right track.

Ever since the publication of Freud's *The Interpretation of Dreams* in 1900, the importance and significance of dreams has been well known. Freud called them the "royal road to the unconscious." I have found that, in addition to their diagnostic value, they are often therapeutically helpful by assisting people to embrace emotions more completely. The case of a man we will call James, with whom I had been working for a year, is a good example.

James was a very bright forty-year-old professional, married and with children. Because of a particularly well defended character structure, blocking certain painful memories, he had experienced only a moderate, intermittent reduction in TMS symptoms. He agreed to the idea of working with dreams. A common theme in many of his dreams was his attempting to leave his parents' house and find his way home. One interpretation of these dreams was that he was trying to establish his emotional independence but was in unconscious conflict about doing so. This interpretation was supported by the fact that his pain began shortly after he started a solo private practice.

One day James told me of a dream that had a very different feel to it. He was chasing a crab that was flying around the room, with the intent of killing it with a knife. He finally succeeded in killing it. He remembered that there was also a cat and a mouse in the room, but they disappeared early in the dream. With regard to TMS, the crab could be a metaphor for the hard, outer shell, symbolically representing pain, other physical symptoms, and/or anxiety that were covering up and protecting him from vulnerable, inner emotions. In addition, he thought that the cat and mouse might symbolize his father and mother.

Usually, in talking to patients during their first meeting with him, when Dr. Sarno asks what their childhoods were like, most people say it was fine. With closer questioning, however, it often turns out not to have been fine at all. So it was with James. Since we had begun to look at his dreams, something seemed to have loosened up in his unconscious and he began to remember things that were not so pleasant. He gradually began to remember much tension and fighting between his father and mother (like a cat and mouse, respectively). I shared with him the idea that the flying crab reminded me of something he had recalled that had happened when he was seven. During one of his parents' fights, his father smashed three plates against a wall, terrorizing James. There was no latent anger or rage against his parents in the crab dream that must have been there because of the tense, often terrifying environment they had created. But James unconsciously did not want to embrace this rage (so the cat and mouse disappear). Instead, he redirects his anger at himself by killing the crab. He must unconsciously also fear that if I continue to penetrate his symbolic, protective TMS shell, he will begin to experience these overwhelmingly painful emotions and they will "kill" him.

These interpretations of James's dream may have resonated with him, but over the previous few weeks his pain level had increased slightly. This is not uncommon in the course of treatment and often can be a positive sign that the person is coming closer to embracing an emotion. The increase in pain is intended to further distract the person from the anticipated greater emotional pain that is considered more dangerous than the physical pain. This is a paradox that is often difficult for patients to grasp: that the physical pain is intended to protect them, not to harm them.

At this point I asked James if he would be willing to experiment with the dream. I suggested he close his eyes, take a series of deep breaths, let go, and just relax. Then I asked him to conjure up the crab in his mind's eye and ask it if it would mind speaking with him and answering a few questions. I encouraged him to continue to breathe deeply, relax, and try to let go of any intellectual attempts at forcing an answer. I told James that if an answer were forthcoming, he would perceive it and there would be no need for him to consciously create it himself. He was immediately pleasantly surprised as the crab had said yes to his request. I then suggested that his crab should be considered a very wise part of his unconscious mindbody and might provide helpful answers to questions he might have about himself. As silly as it might sound, I encouraged James to apologize to the crab for having killed it "to improve the relationship." The reason for this was to plant the seeds for self-compassion, for James, like most people with TMS, tended to be very critical and tough on himself.

Though his pain varied in severity, James was feeling stuck that it was not going away. Now that the crab was willing to dialogue, I suggested that he ask what it thought was getting in the way of his ability to embrace more emotions and reduce the TMS pain. He reported that the crab said it was reluctant to "open the floodgates" for fear of being overpowered by the emotions. I told him to assure the crab that he could now handle whatever emotions surfaced, something he could not do as a child when the painful emotions had to be repressed so he would not be overwhelmed by them. Part of James's conflict as an only child was the need to become emotionally and psychologically independent while simultaneously feeling guilty that he was abandoning his needy parents. That is why his symptoms began when he started his own private practice. This is an example of how "good" things in a person's life can precipitate physical symptoms when there is an underlying conflict. By dialoguing with the crab James began to allow himself over time to embrace the emotions of terror, guilt, and anger at his parents for having created such a terrible atmosphere at home when he was a child.

As the process continued James became aware, to his surprise, of an even deeper level of emotion—that submerged beneath the anger were sadness, sorrow, and hurt for himself. This reflected the beginnings of self-compassion and thus healing through reintegration of mindbody. The anger, at times, is designed to protect the person from these emotions since they are considered more painful than the anger. While experiencing this sadness and sorrow, James reported some reduction in the level of tension and pain. Still not satisfied with his progress, at my suggestion he asked the crab what he should do next and got the immediate response, "Don't be lazy or afraid." This referred to James's reluctance to continue the process at home for fear of the painful emotions he would experience. At the next session he reported momentous progress. While dialoguing with the crab, he was able to experience rage at his parents, which would have been impossible when he was a child. That led him in turn to embrace the sadness and sorrow more deeply, a difficult and painful experience, but one that led to a dramatic reduction of his physical symptoms.

In 1918, in the *Journal of Mental Science* Henry Maudsley, a pioneer psychiatrist said: "The sorrow which has no vent in tears may make other organs weep."

The ability to embrace sadness, hurt, or sorrow for oneself signifies a letting go of the self-critical aspect of one's personality and the development of self-compassion, which is a crucial ingredient for the successful reduction of psychosomatic symptoms.

Although the art and science of psychotherapy includes elements that transcend specific techniques, by utilizing the concepts and techniques described above as well as a variety of others in psychotherapy for people with psychosomatic disorders, therapists help them to observe themselves in the act of repression and become more in touch with emotions like fear, anxiety, guilt, anger, rage, shame, hurt, sadness, sorrow, perceived undeserved joy, and so on. At first, these emotions are experienced as unpleasant, uncomfortable, and, by definition, especially painful. As patients become more comfortable with the discomfort of embracing these emotions, they "come to their senses"; that is, they reintegrate mindbody by allowing themselves to let go of that tight, rigid grip the unconscious mindbody has on them in the form of physical symptoms. By loosening this grip the mindbody no longer has to cause other organs to weep.

AN OUTCOME STUDY

In chapter 3, we made reference to a cohort of 104 patients on whom data were collected to determine the most important psychodynamics at work in cases of TMS. These patients were seen consecutively over a period of two and a half months during the summer of 1999. The following spring we were able to reach eighty-five of the group for the purpose of determining the outcome of the treatment program. They were interviewed on the telephone between six and seven months after the initial consultation.

There were thirty-three (39 percent) males and fifty-two (61 percent) females in our group. Each had participated in one of four different treatment categories:

1. Consultation and lectures only: fifty-nine patients (69 percent)

2. Consultation, lectures, and group meetings: five patients (6 percent)

3. Consultation, lectures, group meetings, and psychotherapy: twelve patients (14 percent)

4. Consultation, lectures, and psychotherapy: nine patients (11 percent)

We were interested in outcome in terms of both level of pain and functional capacity.

The categories for level of pain were as follows:

1. Thirty-seven patients (44 percent) reported they now had little or no pain.

2. Twenty-two patients (26 percent) reported they were now 80 to 100 percent improved.

3. Thirteen patients (15 percent) reported they were now 40 to 80 percent improved.

4. Thirteen patients (15 percent) reported no change to 40 percent improvement.

The categories for functional gains were as follows:

1. Forty-six patients (54 percent) reported they were now unrestricted physically.

2. Eighteen patients (21 percent) reported they were 80 to 100 percent of normal.

3. Twelve patients (14 percent) reported they were 40 to 80 percent of normal.

4. Nine patients (11 percent) reported no improvement to 40 percent of normal.

These figures are extraordinary when one considers that the treatment of this physical disorder is *educational*, augmented in some cases by analytically oriented psychotherapy. Seventy percent of this group had "good relief from pain" and 75 percent were "restored to normal or near normal physical function."

In any treatment regimen one must be aware of the problem of nonconformity—patients who do not take their medications, follow instructions, and the like. With TMS treatment some patients do not follow the daily study program consistently or at all. Others do not call to report their lack of progress so that additional therapeutic measures can be instituted. There are no doubt multiple reasons for this failure to comply, particularly with this disorder in which complicated emotional factors are at play. Though the screening procedure has proven to be quite successful, we may fail with patients who have second thoughts after starting the program and decide not to continue. Others continue to have confidence in the program but are unable to abolish the idea that the structural abnormality blamed for the pain, such as a herniated disc, is still causing the pain.

The phenomenon of psychosomatic symptomatology gives us a psychodiagnostic tool of great power. One can follow the progress of someone with pneumonia by observing fever, cough, respiratory rate, x-ray, blood count, and so on. When a patient states that he is finding it difficult to believe that his pain is emotionally induced, I know that there are strong intrapsychic forces feeding that denial. Since the purpose of symptoms is to keep attention focused on the body, if the patient can be convinced to ignore physical symptoms and focus instead on psychological matters, the psyche's strategy will have been defeated. Therefore, the psyche will work hard to promote disbelief and maintain the status quo. The persistence of symptoms is like the persistence of fever. It gives you some idea whether or not you are being successful in your treatment.

Patients are routinely advised that they need not strive to change their personalities in order to be successful, since so much is made of the perfectionist and goodist traits as well as other problematic personality characteristics like feelings of inferiority, narcissism, and dependency. One cannot change one's inherent traits, though they can be consciously modified. For example, the person who feels perpetually compelled to make a doormat of himself and do nice things for others can take stock and decide to curb the tendency.

Another important therapeutic reality emphasized in the lecture is that the unconscious will resist change so that one must work on the program consistently, and be patient. It was beautifully put by Edna St. Vincent Millay:

Pity me that the heart is slow to learn
What the swift mind beholds at every turn

Not uncommonly, patients develop entirely new symptoms that are part of TMS but do not remember my admonition at the end of the lecture to call me should that occur. Since the symptoms are unlike anything they have experienced before (this is the symptom-imperative at work) they do not think of TMS. Occasionally, the result has been unnecessary surgery.

An example: the patient was a woman who had been successfully treated previously for back pain. She called and said that pain in the right shoulder had been diagnosed as a torn rotator cuff; surgery was performed, with relief of the pain (excellent placebo). When the same pain began in the opposite shoulder a few weeks later her suspicion was aroused; she called and asked if it could be TMS. I told her that TMS tendon pain in the shoulder was often mistakenly attributed to a torn rotator cuff (whose existence is confirmed by MRI) and arranged to see her in the office. When she came in a few days later she said the pain had gone after our telephone conversation. Examination disclosed tenderness of a tendon on palpation. She had TMS tendonalgia at the left shoulder. It is likely that that was the cause of the right shoulder pain.

I diagnose TMS as a physician; I treat it as a teacher. Patients must be educated and inspired. I tell them, "You have a secret weapon—your brain. It may be the instrument of your physical symptoms, but it's also the means by which those symptoms can be abolished."

FIVE

HYPERTENSION AND THE MINDBODY CONNECTION: A NEW PARADIGM

Samuel J. Mann, M.D.

Samuel J. Mann, M.D., a physician and researcher, is an associate professor of clinical medicine at the renowned Hypertension Center of the New York Presbyterian Hospital–Weill/Cornell Medical Center. His work, which encompasses both medical and psychological aspects of hypertension, has been featured in the *New York Times* and other publications. He lectures widely and has published many articles in professional journals. He is the author of a book on the mind/body connection of hypertension, *Healing Hypertension: A Revolutionary New Approach* (Wiley, 1999).

Dr. Mann's chapter is included in this book because, to my knowledge, he is the only expert in hypertension who has established the fact that a significant number of people have high blood pressure due to repressed, unconscious emotions.

HYPERTENSION AND THE MINDBODY CONNECTION: A NEW PARADIGM

Are psychological factors a major cause of hypertension? Many people think so, many others do not. I believe that in some cases there clearly is a mindbody link, although in most cases there is not. I also believe the mindbody link is very different from that which most people assume.

There is no single cause for hypertension, but instead, a mosaic of causes. Studies show that up to 40 percent of hypertension is determined genetically, and up to 30 or 40 percent is determined by lifestyle factors such as diet, weight, salt intake, lack of exercise, and alcohol abuse. I believe that psychological factors provide an explanation for about 20 to 25 percent of hypertension.

In this chapter, which is a brief version of my book, *Healing Hypertension: A Revolutionary New Approach* (1999), I hope to convey an understanding that differs substantially from the usual mindbody theories. I will also emphasize the importance of identifying whether an individual's hypertension is or isn't related to psychological factors, because the choice of treatment depends on this distinction.

I will refer to an old and a new paradigm of the mindbody link in hypertension. The old paradigm is the long-standing, widely believed view that the emotional distress we feel leads to hypertension. This view has regrettably dominated psychosomatic research, without improving our understanding or treatment of hypertension. I hope to show that a new paradigm, which focuses on the emotions we repress and are unaware of, makes eminent sense and, unlike the old paradigm, can have a major impact on the treatment of hypertension and of many other conditions that are also poorly explained by the old paradigm.

THE OLD AND NEW PARADIGMS

The old popular paradigm is that people who tend to be tense or angry, or who face a lot of day-to-day stress, are at increased risk of developing hypertension. Four decades of psychosomatic (mindbody) research have sought to prove this view and to prove that stress-reduction techniques can alleviate or prevent hypertension. They have failed. I believe the shackles of this old, tired point of view need to be released before we can better understand and treat hypertension and other disorders whose mindbody link we suspect but cannot grasp.

I will present the very convincing evidence that this old paradigm is wrong, and will present a new paradigm that makes more sense of hypertension by linking hypertension to the emotions we repress. This paradigm requires recognition of the role of the unconscious and of the important physical effects of emotions we do not feel or realize we harbor. This understanding can remove the mystery of the mindbody connection of hypertension and many other disorders and lead to better treatment approaches.

THE OLD PARADIGM

People who tend to be tense or angry, or experience considerable day-to-day stress, experience repeated elevation of blood pressure and ultimately develop hypertension. Techniques that reduce stress can ameliorate hypertension.

This view suggests that people who tend to be anxious or angry are more hypertension prone than others.

Marie is a worrier. She came to see me because she was very worried about her hypertension. She also believed that she was causing her hypertension because she worries about everything. Sure enough, her blood pressure was elevated in my office.

Marie is the kind of patient that psychologists who adhere to the old paradigm would regard as the classic hypertensive personality. They would teach her relaxation techniques as holistic treatment for her hypertension. If she could learn to relax, to worry less, her hypertension would lessen and she might be able to avoid or get off medication.

I see many patients like Marie. I view her very differently, and the first thing I would tell her is that her worrying is not the cause of her hypertension. The second thing I would tell her is that it is possible she doesn't even have hypertension, and if she does, it is probably because of a genetic predisposition and not because of her worrying.

Most people who believe that there is a mindbody connection in hypertension focus on consciously experienced emotional distress as the link. I do not believe in this simplistic notion, based on what both the studies and clinical experience tell me.

The old paradigm is very simple and would provide the following explanation for Marie's hypertension: *Stress and emotional distress raise our blood pressure. Repeated encounters with stress repeatedly elevate our blood pressure, which ultimately damages, thickens, and stiffens our arteries, leading to persistent blood pressure elevation, or hypertension. The emotional distress may be related to external events or may be internally generated even in the absence of major stressors.*

According to this view, if we can learn to handle stress in a better way, we have a better chance of not becoming hypertensive. If long-standing tension or anxiety or anger causes repeated blood pressure elevation that ultimately leads to hypertension, then relaxation techniques and anger management can lower our blood pressure reaction to stress and prevent hypertension. Thus, we have a complete loop of cause and prevention or treatment.

This is a neat package, with only one problem. It is very wrong. It is intuitively attractive but wrong, and decades of studies tell us it is wrong.

The only point in this old paradigm that holds up in studies is that when we get angry or anxious, our blood pressure does increase, sometimes substantially. However, this response is temporary. It doesn't persist. If you get angry at your spouse, regardless of who is at fault, and regardless of whether you have hypertension, your blood pressure will likely rise and then come down. This will happen again and again and again, and has nothing to do with whether you will ultimately develop hypertension. This is a normal physiologic reaction. Your spouse is not the cause of your hypertension.

It is also clear that if you run to first base, ride a bicycle, or carry heavy packages, your blood pressure will increase in the moment. And no, running to first base does not increase your risk of developing hypertension.

Many people do not realize that it is normal for our blood pressure to fluctuate and that these fluctuations do not cause damage or lead to disease. I can guarantee that your blood pressure will be lower if you sit in a chair all day than if you are active and interact with people. I can also guarantee that sitting in that chair with a lower blood pressure will not prevent development of hypertension.

When a concept contains truth, research ultimately hones in on that truth, even if there are bumps and false starts along the way.

However, the psychological studies attempting to prove the old paradigm are mired in bumps and dead ends. The end result of billions of dollars and decades of this research is a quagmire of conflicting results that have failed to confirm the paradigm or to impact our understanding or treatment of hypertension. Despite this, researchers continue to design and get funding for yet more studies.

Emotions that we feel clearly can have physical effects, such as a temporary increase in heart rate or blood pressure, a tension headache, diarrhea, and other effects. Blood pressure fluctuation in response to stress is simply another transient physical effect of conscious emotion. It does not lead to sustained hypertension.

My clinical experience tells me this every day. Every day a patient warns me that his blood pressure will be elevated because of office stress or horrendous traffic. I know from experience that the traffic jam will not affect his blood pressure in my office. Yes, his blood pressure may have been higher while he was sitting and cursing in his car, but the effect is long gone by the time I measure it. Day-to-day stress does raise our blood pressure in the moment, and it can make us miserable, but it does not lead to hypertension.

WHAT THE STUDIES SHOW

I would like to briefly summarize the results of studies that have examined the relationship between stress and hypertension. I will focus on a few of the most pertinent questions:

1. Are people who tend to be angry or anxious prone to develop hypertension?

2. Since most people spend the lion's share of their time on the job, does job stress lead to hypertension?

3. Do stress management techniques alleviate or prevent hypertension?

To be honest, for any of these questions, I could provide evidence for any point of view. That is part of the problem with psychosomatic research in hypertension: there is no consistency in the findings. I will therefore focus mostly on reviews rather than on individual studies, except to illustrate certain points.

The following key questions pertain to the old paradigm:

- Does the tendency to be anxious, angry, or tense cause hypertension?
- Does job stress cause hypertension?
- Do stress management techniques alleviate or prevent hypertension?

Hundreds of studies focused on people with borderline or mild hypertension have examined this question. If a relationship between hypertension and emotions such as anger and anxiety existed, one would think the evidence would be clear by now. Instead, results vary tremendously from study to study, offering support for any and every point of view. Reviews have concluded that if there is any link between either anxiety or anger and hypertension, it is very weak (Suls, Wan, and Costa, 1995; Jorgensen et al., 1996). Even in people with the most severe hypertension, anger and anxiety scores were no higher than in people with normal blood pressure (Mann and James, 1998).

Studies of anger and hypertension have focused largely on two forms of anger, "anger-out" and "anger-in." Anger-out is the anger we feel and express. The guy with a high anger-out score is the type of person who frequently explodes in anger. Everyone knows he is angry. The studies are clear that this type of person is not prone to develop hypertension (Jorgensen et al., 1996).

The other type of anger, anger-in, might be a different story. Study results are inconsistent, but they do suggest that people who hold their anger in are slightly more prone to develop hypertension than people who get it out (Jorgensen et al., 1996). Here as well, though, the relationship is not a strong one.

There is also ambiguity about the meaning of "anger-in." Many consider anger-in to be the tendency to feel anger but hold it in and not express it. However, the nonexpression of anger is not limited to this pattern. For example, some people get angry but decide that the provocation is not worth the anger and walk away and forget it. This would seem to be a very healthy way of handling anger. Others stew inside but say nothing. This may be closest to the conception of anger-in explored by research psychologists. Still others just don't feel angry. They are not holding in anger that they feel—they simply don't feel angry. These are clearly very different ways of handling anger—holding in anger we feel versus not even feeling it. Unfortunately, it is often unclear which of these forms of anger is being assessed in a given study.

If emotional distress does not lead to hypertension, then what has created the myth that tense people are hypertension prone? One source of this myth is that emotions such as anxiety and anger clearly do raise our blood pressure, although only in the moment. Another is that anxiety in the doctor's office raises blood pressure when a doctor is measuring it, incorrectly giving tense people a diagnosis of hypertension even though their blood pressure is usually normal.

My clinical experience is consistent with what the studies show. I see many patients who tend to be angry or tense, and I do not find them to be more hypertension prone than anyone else. They might, if anything, be more prone to white coat hypertension—blood pressure elevation in the doctor's office with normal readings at home.

I see many patients who are worried about their blood pressure and have elevated readings when I measure it. They see themselves as having worry-induced hypertension, when in fact many do not really have hypertension. Similarly, many patients check their blood pressure at home specifically when they are upset and, sure enough, obtain elevated readings. They conclude that anxiety is making them hypertensive, not realizing that these readings are not representative of their usual blood pressure.

The story of a patient of mine dramatically illustrates the absence of long-term effects of emotional distress on blood pressure. The case provides the kind of powerful observations from the real world that formal studies cannot provide.

I had been following Susan, fifty-six, for borderline hypertension for two years. At a routine follow-up visit, she informed me that her son, thirty-two, had bean diagnosed with an advanced form of malignant melanoma and was likely to die. During the following year, I had the opportunity to observe the response of her blood pressure to this horrible stress, a stress infinitely more severe than the mild stress applied in stress research in hypertension.

Her son died a year later. During that year, Susan was upset and angry and frightened and distressed to the limit. She frequently cried in my office. Yet her blood pressure did not budge a millimeter during that year.

Studies of stress and hypertension often look at the very brief reaction of blood pressure to typically mild stressors, such as mental arithmetic or tracing of an object using a mirror image, or the stress of simulated public speaking. I don't mean to minimize the stress of mental arithmetic, but these stresses pale in the face of both the severity and duration of the stress Susan endured, a stress that cannot be simulated in an experimental situation. The case illustrates how the transient blood pressure increase seen in response to minor laboratory stress, which occurs again and again in response to day-to-day stress in real life, has nothing to do with long-term blood pressure responses to major stress.

Patients like Susan have taught me that even severe emotional distress does not lead to sustained blood pressure elevation. This is a statement that flies in the face of what most people believe, but it is a cornerstone observation that requires us to look for a different paradigm for the mind/body connection in hypertension.

DOES JOB STRESS CAUSE HYPERTENSION?

People spend more waking hours at their job than anywhere else, and, if stress causes hypertension, I would expect job stress to be high on the list of causes. Clearly, in some indirect ways, job stress can lead to higher blood pressure, if it leads to alcohol abuse, overeating, weight gain, or insomnia, factors that are known to cause elevation of blood pressure. But does job stress directly cause hypertension?

Here again, my experience and the results of studies are in agreement that it does not. I see many patients who are experiencing considerable work stress, for example, long work hours, a boss who is overly demanding, or other problems. My experience is that it is the rare patient who experiences a long-term effect on his blood pressure. And here as well, the studies—many of which were performed by researchers who were convinced that job stress causes sustained blood pressure elevation and hypertension, and whose task of getting grants depended on providing evidence to support this view—failed to prove that this widely held belief is true.

I have searched the literature and found forty-eight studies that address this question. As in other areas of psychological research in hypertension, the results are very inconsistent; in the aggregate, they do not support the contention that job stress, per se, leads to hypertension.

The best study ever performed to investigate this issue, was reported by Fauvel (Fauvel et al., 2003), who assessed blood pressure in 209 subjects, using a twenty-four-hour ambulatory blood pressure monitor, at the beginning and end of a five-year period. This study found no relationship whatsoever between job stress and blood pressure.

Researchers who are committed to finding a relationship between job stress and hypertension, and for whom future grants depend on finding it, will almost always find something, anything, that enables them to publish a positive report. When we look at this research for what it is, the evidence is weak and inconsistent. The findings in Fauvel's study appear definitive, and I hope they will put an end to the futile performance of yet more studies.

DO STRESS MANAGEMENT TECHNIQUES ALLEVIATE OR PREVENT HYPERTENSION?

Many studies have attempted to prove that relaxation techniques lower blood pressure. Again, the results are all over the lot. A major review concluded that relaxation techniques were of minimal value in treating hypertension (Eisenberg et al., 1993). That review noted that studies that did show an effect tended to lack a comparison control group.

There is no doubt that in the moment the relaxation response, as popularized by Benson, does lower blood pressure. However, these techniques do not appear to provide a sustained effect. The one possible exception was a well-controlled trial that reported that transcendental meditation lowered home blood pressure by 10 millimeters (Alexander et al., 1996). However, in a more recent study by the same researchers, the observed effect on blood pressure was much weaker (Schneider et al., 2005).

WORRYING AND BLOOD PRESSURE ELEVATION: MOVING AWAY FROM THE OLD PARADIGM

Before moving on to the new paradigm, I first want to reiterate a harmful misconception promoted by the old paradigm. It bears repeating that worriers who worry about their blood pressure are not causing their hypertension. In many cases, their worrying is elevating their blood pressure while it is being measured, even if their blood pressure is otherwise normal. The consequence, though, is that the high reading leads to treatment that in many cases is not necessary.

In my experience, worriers often have "white coat hypertension" rather than true essential hypertension. That's why it is essential to measure the blood pressure outside the doctor's office, with an easy-to-use digital home blood pressure monitor or an ambulatory monitor worn for twenty-four hours.

Many doctors instill the fear of stroke in their patients, feeding both the patient's fear and the misleading elevation of the blood pressure at the time of its measurement. This is where a dash of reassurance goes a long way. When I suggest to a worried patient with a high reading that she, in fact, might not have hypertension, the surprise and relief often results in lower readings, sometimes immediately and sometimes after a few weeks. I am not curing her hypertension because she doesn't really have hypertension in the first place. This is perfectly analogous to telling a patient with back pain that there is nothing wrong with his back. That knowledge can have an immediate and profound effect.

I also find, ironically, that people with the mildest hypertension seem to worry more about it than those with severe hypertension, rather than the other way around. Although concern about hypertension is appropriate and helpful, extreme worry about mild hypertension is inappropriate since the risk of stroke is low, may take effect only after many years or decades, and can be minimized with treatment. Sometimes excessive worrying about mild hypertension reflects a tendency to worry about everything. Or the worrying may divert the person's attention from other issues, as back pain often does.

Should worriers with elevated blood pressure consider psychotherapy? Not if the goal is treatment of hypertension. If the worrying affects quality of life or is diverting attention from other issues, then psychotherapy or relaxation or stress-reduction or anger management techniques my have a role. These interventions are useful for controlling tension, not hypertension.

THE NEW PARADIGM

It is not the emotional distress that we feel, but those emotions we have repressed and are unaware of, that leads to hypertension. The process by which we unknowingly keep distressful and threatening emotions from awareness causes persisting stimulation of the sympathetic nervous system (SNS), resulting in persisting elevation of blood pressure. A shift in our awareness, or use of antihypertensive drugs that block the effects of the SNS on blood pressure, or, in the case of episodic hypertension, drugs that help maintain repression, provide the best results in treating this type of hypertension.

The new paradigm is virtually the opposite of the old. Instead of focusing on the distressful emotions people feel, it focuses on the tendency to repress rather than feel painful emotions. The emotions we feel make us miserable and elevate our blood pressure in the moment but do not have a long-term impact on our blood pressure. Instead, it is the lack of awareness of distressful emotions that marks our susceptibility to developing hypertension.

A key principle of this paradigm is that it is not day-to-day stress that causes hypertension. It is instead that which has been repressed, which, without conscious perception or outlet, persists within us for a lifetime, and produces long-standing rather than transient stimulation of the SNS, leading ultimately to sustained blood pressure elevation. Viewing the emotional distress we feel as a prelude to hypertension has not provided the answer to the mind-body link of hypertension. Looking instead at what is hidden beneath a calm veneer tells us much more.

The emotions we repress are often much more powerful and painful than the emotions we consciously experience. These emotions make much better sense as the link to hypertension both because they are more powerful than the emotions we experience and because, unlike the emotions we feel, we never rid ourselves of that which we have repressed. Yet everyone—from patient to physician to research psychologist—focuses on the milder distressful emotions that we feel and report as the cause of hypertension, without seeing the role of the much more powerful emotions repressed within us.

With the mountain of evidence arguing against it, the old paradigm should have been discarded a long time ago. However, research psychologists and most other people won't let it go because the belief that emotional distress causes hypertension is very widely rooted and because there is a widespread lack of recognition of the role of the unconscious.

The principles of the new paradigm are dramatically illustrated in the case of a patient I saw two or three years ago.

Jim, forty-four, was a tall, slim, very successful, and recently married African-American man. Jim had everything going for him, but was tragically diagnosed with incurable metastatic cancer. He was referred to me because he had recently developed moderately severe hypertension, even though he had lost thirty pounds, which if anything should have lowered his blood pressure. With no other apparent cause, it would seem logical to blame his hypertension on severe distress concerning his diagnosis and poor prognosis. However, and I will never forget his answer, when I asked him if he was very upset he responded, seriously and not sarcastically, "No, I'm not upset. Why should I be upset?" He truly was not upset!

This is a classic example of denial—denial of emotions that might be too painful to bear consciously. If Jim had been distraught, everyone adhering to the old mindbody paradigm would have readily attributed his hypertension to the anxiety. The old paradigm, however, does not make sense in this case. The new paradigm, which focuses on repression and lack of awareness of emotion, does. In fact, no other explanation, medical or psychological, can make sense of this case.

Jim's case serves as a dramatic counterpoint to Susan's case discussed earlier in the chapter. Susan coped very painfully with her son's fatal cancer without any increase in her blood pressure. Jim was calm, but his blood pressure climbed and remained elevated. Together, these two examples illustrate the important reciprocal relationship I see repeatedly between the emotional and autonomic reactions to chronic stress.

People who tend to experience severe emotional distress in severely stressful life circumstances are much less likely to develop hypertension than people who repress the emotion and are not distressed.

Those who repress emotions related to overwhelming stress or trauma, past or present, often cope very well because they are not paralyzed by emotional distress. Repression enables us to move on and function in the aftermath of severe trauma. However, without sooner or later confronting the emotion consciously, we are more likely to ultimately suffer psychological consequences or physical consequence such as hypertension or other poorly understood psychosomatic conditions.

A very telling example that again illustrates the lack of relationship between emotional distress and hypertension, and is consistent with the opposite paradigm that links hypertension to repressed emotion, is the dramatic condition known as malignant essential hypertension. This severest form of essential hypertension, if left untreated, will result in stroke, kidney failure, or death within a year, even in people in their forties. Decades of studies have failed to uncover a cause.

If emotional distress caused hypertension, it would seem likely to be more evident in people with severe and unequivocal hypertension than in those with borderline or mild hypertension, yet the opposite seems evident. People with malignant essential hypertension almost invariably seem to be, and are, the nicest people one could hope to meet, regardless of the circumstances of their life. They tend not to be angry or anxious or depressed. Research psychologists have not performed formal studies to look at this phenomenon because they focus only on people with mild hypertension.

Malignant essential hypertension is largely a disease of African Americans, mostly those raised amid the stress and trauma of ghetto life. Most have ample reason to be angry or depressed, yet act as if they are on top of the world. Physicians and psychologists who observe this conclude, given the absence of emotional distress, that there is no mindbody connection in malignant essential hypertension. Ironically, it is the same lack of anger or depression that convinces me that there is.

When they acknowledge to me a history of childhood abuse or trauma in the ghetto, and I see their contented countenances, I know they are repressors. They have survived psychologically by repressing emotions that would otherwise have tormented them the rest of their lives. Repression of emotion is a true blessing to them, but it is contributing to the hypertension.

REPRESSION

We are fortunate in having at our disposal both conscious and unconscious defenses to deal with extremely painful emotions. When we refer to repression, we are talking about the unconscious, something most people don't even think of as affecting our blood pressure.

We repress emotions without knowing we are repressing them.

This is a simple and important statement, but one that many people either do not comprehend or do not believe. Repression does not involve making a conscious effort to put aside emotional pain. We simply don't feel the pain, essentially a gift from our unconscious defenses that keep it from our awareness. This is very different from the conscious defense of diverting our attention away from emotions we know are distressing us, for example, by keeping busy.

An example that illustrates conscious and unconscious defenses is the process of grieving. We could not function during the grieving process if we were overwhelmed by the pain all day every day. We use our defenses, both conscious and unconscious, to shield us from the pain. Sometimes we repress the pain and don't feel it, and are surprised that we feel okay, as if nothing had happened, even though we know that we have been deeply wounded. At other times we feel the pain and make a conscious effort to keep it out of mind. We keep busy and focus our attention on the tasks at hand to avoid thinking about our loss. And, of course, there are other times when the pain scars us. These are the times when we actually are doing the work of healing, by feeling the pain until it eventually begins to ease and becomes more tolerable. We are psychologically healthiest when we can use conscious and unconscious defenses to tamp down emotions that are too much for us, enabling us to encounter these emotions at a pace we can handle, deal with them, and ultimately heal.

In handling emotions related to severe trauma, repression is a very valuable defense for us. I have seen many patients who were survivors of severe abuse or trauma. Many have suffered psychologically the rest of their lives. Others escaped those consequences by repressing the emotion, and seem to be doing much better psychologically than those who did not repress or in whom repression broke down. Repression can lead ultimately to consequences such as hypertension, but the hypertension is treatable and, I would think, preferable to lifelong emotional torment.

In handling emotions related to trauma, conscious and unconscious defenses do not work in isolation from each other. Even people who have repressed all their trauma-related emotions have at one time felt deep pain, whether for days or weeks or longer. In many people, the pain subsides after the conscious handling of a severely painful emotion, leaving a residue of painful but tolerable feelings that come up from time to time. In others, trauma-related emotions are completely repressed and, although the story is remembered, there is no conscious lingering pain.

REPRESSION THAT LEADS TO HYPERTENSION

We all repress. Aside from trauma, we are bombarded with so much that would provoke emotional attention that it would be hard to live our lives without repressing some of it.

Then why don't we all have hypertension? Is repression that is linked to hypertension different than from the repression that everyone utilizes? In my experience, it is. I see two patterns of repression that seem related to hypertension: in some, repression is related to a history of unusually severe abuse or trauma. In others, even without prior abuse or trauma, repression is overused as the main coping mechanism for handling day-to-day stress.

Repression Related to Severe Abuse or Trauma

I have seen many patients who have repressed emotions related to prior severe abuse or trauma, from as long ago as childhood. The fact that the trauma happened so long ago makes it seem unlikely to patients and doctors that it could still be having an effect, particularly since the distressful emotions related to it ceased long ago.

Without repression, a trauma survivor would be at risk of longstanding psychological problems. Many ultimately recover and move on with their lives, but the pain never truly ends. Others don't recover well and may be greatly troubled for the rest of their lives.

In contrast, trauma survivors who have completely repressed these painful emotions report that they have no lingering psychological effects, no matter how severe the trauma. They insist that they put it behind them and moved on, which is in fact what they did. They truly differ from people who have not eliminated all conscious emotion related to those events and who readily acknowledge its unending effects.

One patient who vividly conveyed to me the connection of old trauma to hypertension was Anna.

At forty-nine, Anna had severe hypertension that even five medications could not control.

She had been seeing me for five years when she began to suffer recurring nightmares about being attacked. Seeing me six weeks after the nightmares had begun, her blood pressure was extremely elevated at 250/140. In response to my questions, she acknowledged having been raped by a close relative when she was fourteen, an event she had forgotten about until she recently met the adult son of the rapist. His striking resemblance to his father had triggered the nightmares. After her emotional disclosure, the nightmares ceased and her blood pressure plummeted overnight to 110/80, and remained under control on just two medications. I reported this striking case in the journal *Psychosomatic Medicine* (Mann and Delon, 1995).

When I first began to notice the link between earlier trauma and hypertension, I was under the impression that few people carried a history of truly severe trauma. However, I now realize it is not uncommon, consistent with studies that find a history of severe abuse or trauma in at least 20 percent of people, particularly during childhood (MacMillan et al., 1997). Sadly, research psychologists largely ignore what I suspect is an important link between childhood experience and adult hypertension, focusing instead on day-to-day anger and anxiety.

Many books have been written about the psychological impact of trauma and how to treat it. They focus on survivors who are suffering, with little recognition of the many who have survived without overt psychological consequences such as depression, anxiety disorders, post-traumatic stress disorder, alcoholism, drug abuse, or other problems. These successful survivors have no need to see a psychologist, which is why psychologists rarely write about them. When such an individual sees a physician for uncontrollable or severe hypertension, the physician will not suspect a mindbody link in the absence of overt emotional pain or psychopathology. The answer lies not in the emotions they report, but in their story. But the story does not get the attention that it should.

Repression as a Day-to-Day Coping Style

I see many patients who don't have a past history of abuse or trauma but who are repressors. We all know people who are very even keeled, or who are always up, who rarely get upset, even about the major stuff that gets to most of us. They routinely repress the distressing emotions of day-to-day life.

In my experience, this tendency is associated with hypertension. This may seem counterintuitive, but it is what I have observed repeatedly and consistently. It is saying that someone who never feels depressed, no matter what is going on his life, is more prone to develop hypertension than someone who does feel depressed from time to time.

When I ask a patient if he ever gets depressed or down, a repressor will respond "never," no matter what problems he has had to endure. He is not holding in feelings of depression or anxiety or anger. He truly doesn't feel them.

These observations are supported by the results of many studies. In his meta-analysis of psychological studies, Jorgensen found emotional defensiveness, the tendency to be unaware of emotions, to be the psychological measure most powerfully linked to hypertension (Jorgensen et al., 1996). In a study I published in 1998, I also found this to be true, particularly in subjects with severe hypertension (Mann and James, 1998).

What underlies a person's tendency to handle stress this way? This is not well understood but clearly is a pattern that is already evident in childhood. It is not well studied by psychoanalysts since repressors don't tend to go to psychoanalysts because they are not distressed.

I suspect that, in some instances, a repressive coping style is inborn. That may have been the case with Jim, discussed earlier, who told me that he alone among many siblings was always smiling, no matter how bad things were. He was always the "breath of fresh air" in his family. It was likely a combination of his personality and encouragement by his family, who needed someone to be that way.

In others, a repressive coping style might have resulted from growing up in a family that did not discuss or share feelings or emotional pain, with no one available to comfort them at times of distress. Instead, without that support, they learned, by necessity, to numb themselves, to not feel. Some come from a home where the macho philosophy of not yielding to emotional pain also led to less and less awareness of painful emotion. Yes, a person can be alone even if surrounded by a large family.

Does this mean that being a calm person puts you at risk of developing hypertension? No. Some people are calm and handle day-to-day stress calmly. They "don't sweat the small stuff." That approach to stress does not have to involve repression and is an admirable way to handle the "small stuff." However, when we don't sweat the big stuff either, when emotions related to the big stuff are swept under the rug without conscious effort, that is, when the sympathetic nervous system (SNS) is activated. It is not a bad way to handle overwhelming stress, but we might ultimately pay a price.

DO ALL REPRESSORS DEVELOP HYPERTENSION?

Certainly not. Many repressors have a normal blood pressure. What determines whether a repressor develops hypertension? Two important factors are genetic predisposition and excessive weight. People with one or both of these risk factors are more likely to develop hypertension than people without them. These three risk factors—genetics, overweight, and a burden of repressed emotions—appear additive in leading to hypertension, as I reported in the *Journal of Psychosomatic Research* (Mann and James, 1998).

The reason for this is that our body has many checks and balances designed to keep our blood pressure within a normal range. If one system, such as the SNS, is out of whack, other systems compensate to restore blood pressure to normal. However, if SNS tone is high as a result of psychological factors, and genetic factors such as abnormal salt retention are also raising blood pressure, the combination is more likely to result in abnormally elevated blood pressure.

Another factor that explains why some people who habitually repress don't develop hypertension is life experience. People with a repressive coping style who have encountered unusually severe life stress seem more likely to develop hypertension than those who haven't. In other words, even if you are a repressor, if you have been lucky enough to live a life free of unusually severe stress, you are less likely to develop hypertension than if you have lived a very difficult life.

ARE THERE CLUES THAT IDENTIFY WHOSE HYPERTENSION IS LINKED TO REPRESSED EMOTION?

It is one thing to state that repressed emotions are the link between the psyche and hypertension; it is another to use this information to guide treatment. I believe emotional factors are a driving force in only a minority of people with hypertension. As I will describe, identifying those whose hypertension is psychologically linked is important because different treatment alternatives are available for them.

How, then, can we tell whose hypertension is linked to psychological factors and whose is not? My clinical experience tells me there are clues that we can use. I tend not to suspect psychologically linked hypertension when there are ample other reasons for someone to have hypertension. For example, if someone is overweight, has a strong family history (suggestive of genetic predisposition), and has mild hypertension that responds nicely to a diuretic or an angiotensin-converting enzyme inhibitor (ACEI), this very ordinary case of hypertension is highly likely to be related to genetics and lifestyle. On the other hand, when a patient's personal history suggests repression, or when the pattern of the hypertension differs from the usual pattern, a psychological link is much more likely (table 1).

TABLE 1: Clues to Psychologically Linked Hypertension

1. Personal history
    a. A history of severe abuse or trauma, particularly during childhood
    b. The conviction that severe prior trauma has no lingering effects
    c. A very even-keeled personality, a person who is never "down"

2. Pattern of the hypertension
    a. Severe hypertension
    b. Resistant hypertension

c. Hypertension with a sudden unexplained onset
d. Paroxysmal (episodic) hypertension

PERSONAL HISTORY

A History of Severe Abuse or Trauma, Particularly During Childhood

I suspect that hypertension might be linked to repressed emotions if an individual reports a history of severe trauma. In exploring for such trauma, I always ask about a person's childhood, whether someone had an abusive parent, or experienced the death of a parent, or other traumatic events. I also inquire about events in adult life, such as a history of combat or the sudden loss of a spouse or child.

There are many, many types of trauma, and even when there is a history of severe trauma, we don't always find it; no brief set of questions can cover all bases. Nevertheless, particularly in people with severe or uncontrollable hypertension, such a history can often be uncovered, supporting the link between these forms of hypertension and psychological antecedents.

The Conviction That Severe Prior Trauma Has No Lingering Effects

When a patient reports that prior trauma has had no lingering impact whatsoever, my antennae are up. This is where I suspect that emotions related to the events have been walled off and are contributory to the hypertension. If someone has survived particularly severe trauma without apparent psychological impact, that alone indicates repression, our greatest ally in moving on.

A Very Even-Keeled Personality, a Person Who Is Never "Down"

The patient who tells me he is worried about this or angry about that, or that he or she feels down or depressed from time to time, is not likely to be a repressor. The patient who tells me he is very even keeled, or is always up, or has never ever been depressed no matter what has happened, is likely a repressor.

PATTERN OF THE HYPERTENSION

In most people, hypertension follows a fairly typical course. There is usually a family history of hypertension: the hypertension begins gradually, with borderline readings, or readings that fluctuate between normal and high until eventually higher readings appear. Blood pressure readings are usually below 160/110. Higher readings can, of course, be seen at moments of great stress. The hypertension usually responds well to a single medication, often a diuretic, or to a combination of two drugs, such as a diuretic and an ACEI.

When an individual's hypertension differs from this usual pattern, doctors may suspect that the hypertension has a specific cause, such as kidney disease or narrowing of the artery to the kidney, or overproduction of aldosterone or adrenaline or cortisol by the adrenal gland. However, even among people whose hypertension does not follow the usual pattern, a cause is found in fewer than 10 percent, leaving 90 percent or more with the unsatisfactory diagnosis, by default, of essential hypertension, no matter how severe or resistant to treatment.

It is in this group of patients that I suspect a link to psychological factors. It is in this group that the proportion reporting abuse or trauma is high, and in which the calm veneer is more prevalent. It is in this group that I consider treating with drugs directed at the SNS as I will describe below.

Several atypical patterns of the hypertension serve as the clues.

Severe Hypertension

When a patient has severe hypertension, with readings exceeding 180/110, the diagnosis of plain old essential hypertension is troubling. It is in these cases that I look for, and often find, clues to a psychological link.

Resistant Hypertension

Treatment with one or two medications is sufficient to normalize blood pressure in most people. The usual treatment consists of either a diuretic or a drug that antagonizes the renin/angiotensin system—an ACEI or an angiotensin receptor blocker (ARB)—or a combination of the two.

The most common reason for failure to achieve a normal blood pressure with these agents is the use of too low a dose, particularly of a diuretic. When adequate dosage doesn't do the job, I begin to look for other reasons. This is where, in many cases, the mindbody connection comes into play.

Repression does not explain all cases of uncontrolled hypertension, but in my experience many patients with uncontrollable hypertension have clues suggestive of repression.

Hypertension with a Sudden Unexplained Onset

When someone has had a consistently normal blood pressure and suddenly develops severe blood pressure elevation with no obvious explanation, it is not the usual pattern. The genetic tendency is not one that affects an individual suddenly. When hypertension appears in this manner, I search for clues to a mindbody link.

Paroxysmal Hypertension

In my experience, paroxysmal (episodic) hypertension is *almost always* linked to repressed emotions. In this type of hypertension, patients describe episodes that occur suddenly and out of the blue, consisting of a sudden onset of physical symptoms such as headache, shortness of breath, weakness, light-headedness, or sweating, with a sudden increase in blood pressure that can exceed 200/100. The episodes can last a few minutes or many hours and are followed by a period of exhaustion. They can occur daily or once every few days or weeks or months. In between episodes, the blood pressure can be normal or near normal.

The episodes feel horrible. Patients tell me they feel like they are going to die. Many live in fear of the next attack and are afraid to do their normal activities or to go anywhere. Many people with this disorder have to stop working.

This form of hypertension is starkly different from more ordinary "labile" hypertension, in which people experience swings in their blood pressure when they are tense or upset, and are aware that the blood pressure change is related to that distress. The episodic hypertension that I am describing differs in that episodes come at unpredictable times and do not appear linked to emotional distress.

Paroxysmal hypertension always arouses the physician's suspicion of a tumor of the adrenal gland, called a pheochromocytoma (or pheo for short), which secretes adrenaline and noradrenaline and can cause episodic hypertension. Although paroxysmal hypertension fits the textbook symptoms of a pheo, fewer than 2 percent turn out to have the tumor. Hundreds of papers have been written about this rare tumor, with fewer than twenty papers written about the other 98 percent who don't have the tumor. Studies have failed to provide an explanation or approach to treatment for this 98 percent, leaving suffering patients to consult doctor after doctor after doctor.

I have found and reported (Mann, 1999) that almost all people with paroxysmal hypertension either have a history of unusually severe trauma that they insist is not affecting them or exhibit a repressive coping style. Understanding this psychological basis is not merely of academic interest, because it has finally opened the door to successful treatment. In some cases, a shift in awareness, by itself, halts recurrence of attacks. However, in most cases, successful treatment requires medication. Based on an understanding of the disorder's origin in repressed emotion, we now have pharmacologic approaches that control the disorder in most cases, as I will explain below.

TREATMENT IMPLICATIONS OF THE NEW PARADIGM

As we have discussed, psychological factors underlie hypertension in some people but not in others. When they are operative, the mechanism driving the hypertension differs from the mechanisms driving more routine cases of essential hypertension. Identification of psychologically linked hypertension is therefore very important in terms of selecting treatment that matches the cause of the hypertension.

Diet, exercise, and weight control, and, in many people, restriction of salt intake are certainly important measures and can eliminate the need for medication in some and reduce the amount of medication needed in others. On the other hand, relaxation and stress management techniques, as proposed by the old paradigm, although helpful in managing anxiety and anger, are of little benefit in treating hypertension, as recently reviewed (Eisenberg et al., 1993).

PRINCIPLES OF TREATMENT WITH THE NEW PARADIGM

In contrast to the old paradigm, the new paradigm offers profound treatment implications. It offers strategies for selecting drug and nondrug therapies best suited for each individual.

An extremely important question that is rarely asked is whether hypertension that is linked to psychological factors should be treated differently than hypertension that is not related to psychological factors. This question has been the subject of virtually no research and is not even on the radarscope of researchers, probably because medical researchers are not interested in psychology and research psychologists are not interested in antihypertensive medications. Based on my clinical experience, and supported by my studies, I am convinced that the new paradigm can be very helpful in guiding treatment, as I shall discuss in this section.

Nondrug Treatment of Neurogenic Hypertension: Is There a Role for Psychotherapy?

Any discussion of psychological factors and hypertension must lead to an important and widely asked question: can psychotherapy help alleviate hypertension? If I am correct in my understanding, the surprising answer is usually, but not always, no! Here's why.

When I see patients with hypertension who have survived enormous abuse or trauma by repressing the emotions related to it, I suggest to them the possible origin of their hypertension in events from long ago. Some individuals, at mention of those events, often for the first time in decades, experience an almost sudden, and healing, shift in awareness of the impact of their past. In some cases, I have seen hypertension rapidly melt away when this kind of shift occurs, even without psychotherapy. Psychotherapy can then be helpful in dealing with the emotions that arise from this shift in awareness. However, if a patient is uninterested, I do not coerce him to embark on psychotherapy.

Patients whose hypertension is related to repression usually do not complain of emotional distress, do not feel any need for psychotherapy, and are not interested in it. And they are coming to me not because they want to explore old trauma, but because they want to bring their blood pressure under control. Also, if someone has successfully repressed overwhelmingly painful emotions related to childhood abuse or trauma, and has moved on successfully, exploring them in psychodynamic therapy might be the last thing they would want to do, and might be the wrong thing for them to do. Exploring such powerful emotions, given the barrier of repression, would likely be a waste of time and money anyway. The therapy would be unlikely to crack the repression, and if it did, it would have the potential to do harm. Although I am an advocate of recognizing the mindbody relationship in hypertension, I would nevertheless argue against coercing patients whose successful emotional survival is attributable to repression, to seek psychotherapy and attack that repression.

Books written about therapy for trauma survivors are generally not about the untroubled patient who has put the trauma behind him and doesn't feel affected by it. And it would be wrong to generalize from the principles of treatment of people troubled by past trauma to the treatment of those who are not. It might be better to leave the past alone, even if past abuse or trauma is responsible for the hypertension. It is wise to honor a patient's preference to not explore the past.

On the other hand, when repression is failing, when internal alarms are going off without a source that is obvious to us, marked psychological and sometimes physical symptoms can become apparent. This type of patient appears headed for unending trouble. Here, psychotherapy can be extremely important, although even here it comes with no guarantees.

Is hypnotherapy a worthwhile alternative? Perhaps, although here, as well, we just don't know whether bringing repressed material to conscious awareness through hypnotherapy is always safe, particularly in people who are not suffering psychologically. I therefore cannot uncritically recommend it. I hope someday we will know enough about hypnotherapy to know if it is a realistic and safe alternative.

In summary, the catch-22 of psychotherapy in hypertension, from my experience, is that the more aware an individual is of emotional distress, and the more willing he is to pursue psychotherapy, the less likely it is that the psychological issue is the cause of the hypertension. On the other hand, the more a trauma survivor insists that the past has had no lingering impact, indicating that major issues have been repressed, the more likely it is that the hypertension is related to it, but the less likely it is that such an individual will be amenable to, or helped by, psychotherapy.

Drug Therapy of Psychologically Linked Hypertension

Even if psychotherapy is often not an option, the distinguishing of psychologically linked hypertension from other cases of essential hypertension offers important implications in terms of drug therapy. The specific treatment recommendations in this section are based on the principle that in people with psychologically linked hypertension, the mechanism and treatment of the hypertension differ from those in whom hypertension has nothing to do with psychological factors. I will begin by briefly discussing mechanisms of hypertension, and then providing the rationale for basing treatment on these principles, based on my studies and clinical experience.

Mechanisms of Essential Hypertension and Relationship to Selection of Drug Therapy

Many mechanisms of hypertension have been explored, and a thorough review would be beyond the scope of this chapter. I will instead describe three widely investigated and documented mechanisms that are addressed by drugs that are in wide use. These mechanisms are excessive blood volume, excessive activity of the renin-angiotensin system (RAS), and excessive activity of the SNS.

Blood volume

Hypertension that is driven by excessive blood volume is also known as salt-sensitive hypertension. In genetically susceptible individuals, ingested salt is less efficiently excreted by the kidneys, leading to retention of sodium and fluid in the arterial system, and increased calcium ion levels in the smooth muscle cells of arterial walls, which elevate blood pressure. For particularly salt-sensitive people, even a little salt does this. Salt-sensitive hypertension is particularly common among African-Americans and in the elderly. Restricting salt intake can normalize blood pressure in some people with volume-related hypertension, but for many others drug therapy is needed.

For volume-dependent hypertension, diuretics, which increase sodium excretion by the kidneys, and calcium channel blockers, which block calcium entry into smooth muscle cells, are the most effective agents. Other agents, such as the ACEIs, ARBs, and beta blockers, are less effective in this form of essential hypertension.

Renin-angiotensin system (RAS)

The RAS is a complicated system whose hormonal end product, angiotensin II, raises blood pressure by constricting arteries and by stimulating secretion of aldosterone, a salt-retaining hormone. The kidneys produce a hormone called renin, which is an important activator of this system. Secretion of renin is stimulated when the volume of the blood is low, and is suppressed when it is high. That is why the blood renin level is usually low in most people with volume-driven hypertension, and higher when volume is not a factor. ACEIs, such as enalapril (Vasotec), lisinopril (Prinivil, Zestril, and others), which block the production of angiotensin II, and ARBs, such as valsartan (Diovan), irbesartan (Avapro), candasartan (Atacand), and others, that block the receptors that angiotensin II binds to, are most effective in people whose hypertension is mediated by the RAS.

In most cases of ordinary essential hypertension, excessive blood volume and/or effects of the RAS are involved. That is why a diuretic and/or an ACEI or ARB are so effective and are the mainstay of drug treatment. Given alone or in combination, they bring blood pressure under control in the lion's share of cases of ordinary essential hypertension.

Sympathetic nervous system

The SNS links the brain to our heart and arterial system. It has two limbs, the adrenal limb and the sympathetic nerves.

The adrenal limb traverses from the brain, along the spinal cord, to the adrenal gland, stimulating secretion of adrenaline. Adrenaline, in turn, stimulates receptors in the heart and arteries, with the most prominent effect being stimulation of beta receptors in the heart, causing it to beat faster and harder, and increasing the amount of blood pumped, known as the cardiac output. When we are nervous and feel our heart racing, that is the adrenaline flowing. Adrenaline also causes arteries to dilate, through stimulation of beta receptors in the walls of arteries.

The neural limb consists of nerves that traverse the spinal cord and innervate the heart and arterial wall. They stimulate the alpha receptors in the heart and arterial walls, raising blood pressure by increasing heart contractility and by narrowing arteries.

Stimulation of the SNS usually involves activation, to some extent, of both the adrenal and neural limbs. It is the SNS that mediates the transient effects of stress and emotion on our blood pressure. Stressors such as fear or anxiety stimulate primarily the adrenal limb, manifesting with a rapid heart rate and increased cardiac output and systolic blood pressure. (When blood pressure is recorded as, for example, 120/80, the 120 is the systolic pressure, which reflects the pressure generated by the force of the heartbeat, and the 80 is the diastolic, which reflects the pressure in the arterial system in between heartbeats.) Stressors such as weightlifting stimulate mostly the sympathetic limb, increasing both systolic and diastolic blood pressure, with little effect on heart rate. Emotional stimuli (e.g., anger and most other stimuli) tend to stimulate both adrenal and neural limbs.

Stimulation of the SNS also increases sodium retention and stimulates the RAS. Nevertheless, its main effects are those on the heart and arteries.

Hypertension that is driven by the SNS rather than by volume or the RAS is called neurogenic hypertension. In a recent article I summarized neurogenic hypertension and reviewed its causes, which include emotional factors as well as other factors unrelated to emotions (Mann, 2003).

PHYSIOLOGIC MECHANISMS: HOW REPRESSION CAUSES HYPERTENSION

How, then, does repression of emotion fit in with these mechanisms and lead to hypertension? There are relatively few data to answer this question because few researchers who focus on either medical or psychosomatic aspects of hypertension have investigated or even considered the role of repressed emotions.

We do know that the SNS mediates the transient effects of the emotions we feel on our blood pressure. It is a logical candidate to mediate the effects on blood pressure of repressed emotions as well. Just as the emotions we feel in the moment stimulate the SNS and raise blood pressure in the moment, the emotions we repress chronically stimulate the SNS and produce more long-lasting blood pressure changes.

What is it that triggers the SNS in repressors? Is it the emotions that fly beneath the surface of conscious awareness? Or is it the process of repression, the heightened state of alarm unconsciously defending us against the threat of these emotions? This is unclear. But what seems clear is that emotional processes we are not consciously aware of are having a more long-lasting effect on the SNS than are the conscious emotions that distress us.

DRUG TREATMENT OF PSYCHOLOGICALLY LINKED HYPERTENSION

The widely used diuretics, ACEIs, and ARBs target blood volume and the RAS. They do not target the SNS. One could logically deduce that drugs that do target the SNS, such as alpha- and beta-receptor blockers, would be more effective in treating hypertension mediated by the SNS. And, in fact, they are.

The beta blockers block the effects of adrenaline on heart rate and force of contractility, and lower blood pressure particularly in individuals whose hypertension is characterized by a rapid heart rate. Examples of beta blockers include atenolol (Tenormin), metoprolol (Toprol), and others. My favorite is betaxolol (Kerlone), which has a smoother action and, I believe, fewer side effects. The alpha blockers block the constricting effects of the sympathetic nerves on the walls of the arterial system. Examples include doxazosin (Cardura) and terazosin (Hytrin).

Other drugs, such as clonidine, reduce SNS tone through their effects on receptors in the brain, but have very prominent side effects, such as fatigue. I try to avoid them except under duress.

Sometimes a beta blocker will do the job by itself, particularly in people with a rapid heart rate. Beta blockers also combat anxiety by blocking the noticeable physical effects of adrenaline (palpitations and tremulousness). Often, however, a beta blocker by itself will not do the job, and combining it with an alpha blocker works much better.

This combination—a beta blocker with an alpha blocker—is highly effective, as I and others have demonstrated in studies (Holtzman et al., 1998; Mann and Gerber, 2001), and, perhaps more important, as I have seen consistently in clinical experience treating patients with hypertension that appears linked to repressed emotion and cannot be controlled by a diuretic/ACEI or diuretic/ARB combination.

Based on the mechanisms, I would expect the diuretics and ACEIs and ARBs to work very well in cases of ordinary essential hypertension, but less well in psychologically linked hypertension that is mediated by the SNS. That is what I observe treating patients, and is what I observed in a recent study (Mann and Gerber, 2002). In that study, I used a history of abuse or trauma during childhood as a marker for a likely burden of repressed emotion. I found that in subjects without such a history, a diuretic or ACEI normalized blood pressure in 75 percent of cases. In those with such a history, it was controlled in only 25 percent. This is a startling difference and makes sense, given the more prominent role of the SNS in their hypertension. In this study, subjects who reported childhood abuse responded much better to an alpha/beta blocker combination than to the ACEI or diuretic.

My clinical experience is highly consistent with these results. Patients with a personal history or pattern of hypertension suggestive of psychologically linked neurogenic hypertension are less likely than others to achieve a normal blood pressure with a diuretic/ACEI combination, and respond better to an alpha/beta blocker combination. Unfortunately, the latter combination is not used as often as it should be.

In summary, if an individual's hypertension is strictly a matter of genetics and lifestyle, and has nothing to do with psychological factors, I generally employ a diuretic or an ACEI (or ARB), or a combination of the two. Alternatively, I use a calcium channel blocker. These drugs, alone or in combination, should control hypertension in most cases. However, when the hypertension is psychologically linked, patients respond less well to these drugs, and it makes more sense to employ drugs that block the effects of the SNS, such as the alpha and beta blockers.

John's case exemplifies this approach.

John, thirty-five, developed severe hypertension shortly after being diagnosed with AIDS. The combination of an ACEI and a diuretic didn't touch his blood pressure, which remained at 180/120. His doctor, who had sought my opinion, considered the possibility that severe emotional distress related to the diagnosis was causing the hypertension, but rejected it because John was really cool about it and was not at all upset.

It would have been easy, according to the old paradigm, to blame the hypertension on the expected emotional upset caused by the diagnosis of AIDS. However, John insisted he was not upset. The new paradigm tells me exactly the opposite—that it is the nonreaction that is more likely to lead to psychologically mediated neurogenic hypertension, which is resistant to treatment with a diuretic and ACEI. Replacing his drugs with an alpha and beta blocker combination promptly normalized the blood pressure.

LABILE HYPERTENSION

It is normal for blood pressure to fluctuate. No one's blood pressure is perfectly steady. It is also clear that blood pressure can vary with stress, rising in tense moments and then settling. We do not treat those transient elevations because they are part of the normal physiology of blood pressure.

However, in some individuals, blood pressure fluctuation is either very frequent or very extreme. Or it might fluctuate frequently even in the absence of obvious stress. In patients whose hypertension fits this pattern, the blood pressure fluctuations are likely governed by the SNS, which controls minute-to-minute and second-to-second variations in blood pressure. It is likely that emotions, whether consciously experienced or repressed, are driving the excessive fluctuations. In this circumstance, alpha/beta blocker combinations make a lot of sense, and work better than diuretics, ACEIs, and ARBs.

AVOID EXCESSIVE DOSES OF DRUGS THAT ARE UNLIKELY TO WORK

When a diuretic/ACEI combination does not work, it is common practice to increase the dosage of one or both drugs. This is important, since in many individuals hypertension cannot be controlled without using a relatively high dose, particularly of a diuretic.

Higher doses are not routinely used because their use is associated with many adverse metabolic effects such as increases in uric acid, triglycerides, and, in some cases, blood sugar. However, if someone with volume-mediated hypertension does not respond to a low dose of a diuretic, prescribing a higher dose is the right thing to do. In contrast, in those whose hypertension is not mediated by volume, a higher dose should be avoided because it confers a higher risk of adverse effects with little benefit to blood pressure control. This is another reason why recognizing when emotional factors are integral to the hypertension can have an important impact on drug treatment.

MANAGEMENT OF PAROXYSMAL HYPERTENSION

Recognition of the role of repressed emotions provides the first valid explanation for paroxysmal hypertension and the first successful treatment approach (Mann, 1999). People with paroxysmal hypertension usually don't respond to ACEIs and diuretics, and need a different approach. The new paradigm offers three new approaches to treatment, involving drug and nondrug therapies.

One approach offers a cure, involving a shift in awareness of emotions that have been repressed. If a patient can understand that the disorder is linked to repressed emotions, and experiences a shift in awareness, the disorder can literally disappear.

Jill, thirty-five, lived a very affluent and pampered life. She did not have to work, and her husband was happy to provide everything for her. She claimed she was perfectly happy, but was suffering episodes of high blood pressure and rapid heart rate almost daily. Upon careful discussion, and in a very painful way, she acknowledged to herself, for the first time, her deep unhappiness. She finally faced her deep shame and despair over the absence of meaning in her life, without career or purpose. She realized that she felt useless and was deeply ashamed of herself.

Her episodes ceased promptly with this painful realization, without any psychotherapy. She did, though, have the difficult task of dealing with her unhappiness. With the awareness she now had, she was able to begin to make changes in her life that she would not otherwise have made.

Jill's case demonstrates healing of paroxysmal hypertension brought about by emotional awareness. However, most people with paroxysmal hypertension do not experience this shift because of the resistance to awareness of deeply painful emotions. Fortunately, there are other effective treatment options for this disorder, also based on the new paradigm. One is the use of combined alpha/beta blockade to reduce the severity of the blood pressure swings by blocking the effects of the SNS activation that underlies them. Unfortunately, even if blood pressure swings are somewhat mitigated, episodes do continue to recur in most patients.

The other alternative is the use of an antidepressant, which has a dramatic effect in most patients. Within two weeks, most patients experience cessation of attacks and are restored to a normal life. This has truly revolutionized treatment of this disorder, as I have recently reported (Mann, 1999). An antidepressant—either an older agent, such as desipramine, or a selective serotonin reuptake inhibitor (SSRI), such as Zoloft or Paxil or Lexapro—strengthens the barrier against the repressed but threatening emotions. Even though people with paroxysmal hypertension are not depressed, these agents work extremely well. I suspect it is because paroxysmal hypertension shares many features in common with panic disorder, which also responds extremely well to these agents.

The effectiveness of an antidepressant in this hypertensive disorder is perhaps the most blatant evidence that the mind can be involved in hypertension, and that the mind must be considered in assessing patients with difficult-to-control hypertension.

SUMMARY

In looking at the mindbody relationship in hypertension, it is natural to suspect that the emotions we feel, which clearly do affect our blood pressure in the moment, play an important role. However, decades of research have failed to confirm that these emotions lead to hypertension or that techniques to reduce emotional distress prevent hypertension.

If instead we look at repressed emotions, at the emotions that we don't feel or complain about or even know we harbor, we have a very different approach that can finally make sense of the mindbody connection in hypertension, and offer new approaches to treatment. This understanding identifies the individual who feels the least emotion as the most hypertension prone, rather than the other way around. When we realize that the emotions we repress may have more to do with hypertension and other chronic medical conditions than do the emotions we feel, we open the door to new approaches to treatment.

Too many medical conditions remain unexplained, and mindbody research limited to what the conscious mind reports has failed to help us understand their origin. Too little attention has been paid to what the conscious mind cannot report. I hope this will begin to change.

SIX

MY EXPERIENCE WITH TENSION MYOSITIS SYNDROME

Ira Rashbaum, M.D.

Ira Rashbaum, M.D., is a Clinical Associate Professor of Medicine at the New York University School of Medicine and an attending physiatrist at the Rusk Institute of Rehabilitation Medicine in New York City. He began his TMS training in 1992 and has treated numerous patients with TMS since 1993. He published an article with Dr. Sarno titled "Psychosomatic Concepts in Chronic Pain" in the March 2003 supplement to the *Archives of Physical Medicine and Rehabilitation.* He is the Director of Stroke Rehabilitation at Rusk. He is a frequent guest editor for the American Academy of Rehabilitation Medicine's *Annual Medical Education Guide*. He considers his proudest accomplishment, however, to be his marriage to his wonderful wife, Robin, and his role as father to his sons Benjamin and Joshua.

Dr. Rashbaum has many responsibilities at the Rusk Institute of Rehabilitation Medicine so we are grateful that he has found time to diagnose and treat patients with TMS. Aside from Dr. Sarno, he is the only other physician in New York City who is capable of doing this.

MY EXPERIENCE WITH TENSION MYOSITIS SYNDROME

My first exposure to Dr. Sarno's concepts occurred during my second year of training as a resident in physical medicine and rehabilitation at the Rusk Institute of Rehabilitation Medicine in New York City. I entered an elevator one evening to find a large group of people talking, laughing, exclaiming about a lecture they had just heard. It was Dr. Sarno's weekly lecture for new patients, and I heard enough to be intrigued by some of the things they were saying about pain of psychological origin and how it could be helped by learning its true cause. However, it was to be a year before I decided to learn more about what I had heard. I had begun to think about the cases I had seen in which the structural explanation of the pain didn't make sense and others where no diagnosis at all was forthcoming. Even more to the point, I recalled the severe neck pain I developed the night before my bar mitzvah, worsening allergic symptoms at exam time in high school and college, crushing headaches during medical school, and recurring stomach upsets during my internship. I procured Dr. Sarno's books, *Mind Over Back Pain* and *Healing Back Pain*, and decided I was ready to learn more. I had a three-month elective coming up; that is, I could study with any program of my choice. With some trepidation, I approached Dr. Sarno about doing the elective with him, and he agreed. In fact, he was delighted, since no resident had ever expressed interest in his work.

For the next three months I saw patients with him, attended all his lectures and group meetings, and listened with fascination to psychologists' reports. I began to realize that a doctor had to be more than an engineer to the body, that feelings and who we are had an enormous amount to do with human illness. I was exhilarated by the idea of being a pioneer in this important field of medicine.

At the completion of my training I was asked to remain at the Rusk Institute as attending physician and faculty at the NYU School of Medicine. I decided to accept, since it was a good position and it would give me an opportunity to continue working with Dr. Sarno and sharpen my skill in the diagnosis and treatment of TMS.

I could go into considerable detail about the enormous public health problem posed by chronic pain in the United States and most of the Western world, and of the evidence in the medical literature suggesting that psychological factors are an important reason for the epidemic of chronic pain in the United States. The cost to society is enormous, in the range of $65 to $79 billion annually. The cost in human suffering is incalculable. Rather than summarizing this literature, I would like to share with the reader my experience in working with TMS.

One of my favorite Talmudic quotes is "All beginnings are difficult." This is the story of one of my first patients:

Ms. S was a middle-aged woman with a history of intermittent low back pain for six years. Shortly before I saw her she developed pain in the left leg and feelings of numbness in the left calf extending down to the third and fourth toes, all attributed to bowling. There is an almost universal tendency for patients to blame the onset of pain and other symptoms on some physical activity. The reason for this is explained in chapter 1. She reported being limited in housecleaning, gardening, and her usual exercise program.

Her medical history was important. She had suffered from cardiac palpitations, "sinus" headaches, colitis symptoms, high blood pressure in the doctor's office (white coat hypertension), and dry skin. It is our view that, like TMS, these are psychologically induced; therefore, we refer to them as equivalents of TMS.

Of great social significance is the fact that she had been sexually abused by her brother as a child. She described herself as a perfectionist, that "no one could do a task as well as she could," and that she was a worrier and a people pleaser.

The physical examination disclosed some weakness in the muscles that elevate the front of the left foot (mild foot drop) and decreased ability to perceive a painful stimulus (pinprick) in the front of the leg. There was also marked tenderness when pressure was applied to muscles in both buttocks and the low back, an almost universal finding in people with TMS. The pain, weakness, and sensory disturbance were all due to the fact that spinal nerves were mildly oxygen deprived after they emerged from the spinal canal, which is what occurs with TMS.

An MRI showed a large herniated disc at the lower end of the spine, which, however, could not have been responsible for the pain because of its location in the spinal canal.

I made the diagnosis of TMS; Ms. S agreed and proved the diagnosis to be correct by returning to normal in three months. I have followed her by telephone over the years and except for occasional mild pain in the wrists or knees (usually associated with family stresses), she has led a normal life. She has discussed her stresses with a social worker from time to time.

When Ms. D, a twenty-seven-year-old single woman, came for

an appointment, she said she had been suffering from pain in the left low back, occasionally on the right side, and pain in the left side of the neck, shoulder, and upper back. She said she had experienced back pain since the age of ten, but serious problems began after she graduated from college. The pain gradually worsened and became so severe she could not walk, sit, bend, stretch, or drive a car. She had been unable to work for six months.

In addition to the pain, she suffered from spring allergies, "sinus pressures," headaches with pain behind the eyes, urinary tract infections, vertigo and, most recently, spastic colon. We have found that these are all equivalents, serving the same purpose as TMS. It is probably no coincidence that she had just become engaged when the spastic colon began. During the interview she said she was a perfectionist, a people pleaser, that she avoided confrontation and took things very seriously. There were a number of stresses in her life.

The only findings on physical examination were the three locations of muscle tenderness in the buttocks, low back, and neck—typical of TMS.

Over a three-year period she had consulted two chiropractors, two neurologists, two urologists, a gastroenterologist, endocrinologist, rheumatologist, orthopedist, dermatologist, proctologist, a TMJ (temporomandibular joint) orthodontist, and an acupuncturist. She had every test imaginable, with no diagnosis for the pain. One doctor asked her if she enjoyed going to doctors, and another said he thought she was "faking it." She had resigned herself to living with pain indefinitely and the fact that she would probably not be able to have children because she would not be able to live for nine months without pain medication.

Following the initial consultation, she attended my treatment lectures and then I did not hear from her again until I received a letter a month later. The following is an excerpt from the letter:

> I have been very busy living my new pain-free life; I am so excited. To bring you up-to-date, I am doing GREAT! My life changed dramatically after coming to see you. I have gone back to work, sitting at my desk for hours, and 99 percent pain free. I have been shopping, walking, running, lifting, carrying, driving, sitting, standing, sleeping through the night, dancing, etc. I am now looking forward to having children someday and living the next fifty or more years PAIN FREE. Many people are amazed when I tell them I am pain free. They cannot believe it. You have had the greatest impact on my life. There will always be a special place in our hearts for you and Dr. Sarno.

A FAMILY AFFAIR

A difficult aspect of practicing mindbody medicine is that the majority of people with TMS and equivalent disorders cannot accept the diagnosis. This is particularly frustrating when a successfully treated patient cannot convince family members of the validity of the diagnosis when they, too, have TMS. Nonacceptance appears to be the rule. The following case histories are notable exceptions.

Ms. EH (daughter) and Ms. SH (mother) arranged simultaneous appointments with me for possible TMS. They had read *Healing Back Pain*, and both were well informed about and entirely open to the possibility that they had TMS. Their pain histories and physical examinations were typical of TMS. The major stress in their lives was the death of their father/husband. Both became tearful when this was discussed during my lecture.

Daughter and mother both became pain free and have remained so through the years, with occasional mild TMS or equivalent symptoms that quickly abate after telephone contact or a reexamination.

TREATING THE OLDER TMS PATIENT

I have evaluated and treated a large number of senior citizens on Medicare for more than a decade, with variable success. A Pennsylvania farmer is credited with the folk wisdom, "As we get older we grow more like ourselves." So it is with the elderly person with TMS. Not only are the personality traits that contribute to the need for symptoms more marked, but now he/she has to live with the specter of disability and mortality. Unconsciously, these are enraging, though we may be philosophical about them consciously. Retirement, for both men and women, often leads to a drop in self-esteem, further feeding the rage reservoir, as well as sadness and hopelessness. Other senior citizens who continue to work experience anxiety about their ability to compete with their younger colleagues, or anger at having to deal with bosses who are younger and less capable than they are or who have inherited businesses that they were instrumental in starting. Living with a painful (other than TMS) or disabling condition is common among the elderly. All of these can lead to mindbody symptoms and can be ameliorated by the TMS education program and/or working with a psychologist.

In working with older patients, one often has to distinguish between symptoms caused by legitimate structural abnormalities and those caused by TMS, and treat both accordingly. It is important to bear in mind, however, that symptoms are often erroneously attributed to structural abnormalities that are in reality due to TMS. A good example of this are symptoms blamed on "arthritis," when the "arthritic" changes are merely the result of aging. Contemporary medicine has made a new "disease" of arthritis, hence the monumental sale of pharmaceuticals and a variety of physical treatments.

An example of this diagnostic duality is the gentleman who had

TMS but also had arthritic changes in his knees that resulted in limited range of motion, occasional locking, and difficulty walking. Clearly, one had to treat the TMS as well as the physical disability.

Another patient, a woman with pain in both legs due to TMS spinal nerve involvement, got better but then developed new pains clearly attributable to peripheral vascular disease. Both conditions were appropriately treated.

A third patient was successfully treated for TMS and returned many years later with low back pain and decreased sensation in both feet. I concluded that her back pain was due to TMS and her loss of sensation in her feet was due to diabetic peripheral neuropathy. A surgeon attributed both to spinal stenosis and mild alignment abnormality of a spinal bone and recommended surgery. Her geriatrician was surprised when her back pain went away as a result of her acknowledgment of great rage due to aging. The nerve symptoms improved with appropriate nonsurgical treatment.

Mr. K, a seventy-six-year-old part-time accountant, came for consultation about ten years ago with fifty years of lower back pain. A lumbar MRI revealed an L4–5 bulging disk. His wife had Alzheimer's disease that he acknowledged was a significant stressor. He saw himself as a perfectionist but not a people pleaser. However, he involved himself greatly in the case of his wife since a private caretaker came only four days a week. He accepted the TMS diagnosis, but progress was slow. He was one of the first patients that I referred for psychotherapy. The psychologist's report stated that he was a stereotypical male of his generation who tended to hold his feelings in and thought that crying was a sign of weakness. His pain began to subside when he allowed himself to more fully feel his profound sadness over the psychological loss of his wife while she remained intact physically. Six months after the initial consultation, he felt great, with occasional mild low back pain. He returned to playing golf after many years of inactivity.

Mr. S was also one of the first patients I referred to psychotherapy. When seen at the time of initial consultation, he was sixty-five, a semiretired real estate broker. He had suffered from pain in both buttocks for over a year. As with the majority of my patients, he had had multiple treatments, including physical therapy, massage, acupuncture, steroids, and nonsteroidal anti-inflammatory drugs (NSAIDs), all without relief. (This list includes surgery with many patients.) A lumbar MRI done five months before the consultation showed herniated discs at L4–5 and L5–S.

He said he had identified with some of the psychological factors associated with TMS in the book *Healing Back Pain*. He saw himself as "hyper" and was aware that he tended to suppress conscious anger. He was referred for psychotherapy due to persistent symptoms six months after the initial consultation. The psychologist reported that he felt pressure in his business and, more important, anger and resentment toward his wife to curtail his work hours further. With psychotherapy, he was able to identify these stressors and their impact on his physical well-being. As his psychotherapy progressed, he reported increased feelings of well-being accompanied by longer free periods. He was able to participate fully in his business, made efforts to be more communicative with his wife, and was playing golf again on a regular basis.

MY EXPERIENCE WITH FIBROMYALGIA

In his article on the dilemma of the fibromyalgia syndrome, Dr. Jerome Groopman, a Harvard Professor of Medicine, suggested that the contemporary epidemic of fibromyalgia was analogous to that of neurasthenia, with the Internet and mass media feeding the epidemic based largely on misinformation.

I have managed a number of TMS cases that were diagnosed by other physicians as fibromyalgia. Here are two such cases:

Ms. G was a forty-six-year-old woman who had developed sciatica during pregnancy twenty years prior and had had lower back pain since the birth of her daughter. Several physicians attributed her symptoms to a significant weight gain during her pregnancy. She subsequently developed neck pain that was diagnosed as cervical disc disease as she had numbness in her right index and middle fingers. A prominent neurosurgeon saw no acute indication for neck surgery.

Further questioning revealed that she had a history of sinus infections, bladder infections, earache, TMJ, headache, and irritable bowel syndrome. As noted, these are TMS equivalents. She described herself as a perfectionist and people pleaser who was very responsible. She felt that she had "a thousand pounds on both shoulders" of responsibility during her pregnancy. She saw herself as "the caretaker of the world," raising two daughters on her own and establishing a career. She realized that she "can't say no to people."

The physical examination was remarkable only for multiple areas of tenderness in eleven of the eighteen characteristic fibromyalgia tender points, especially at the top of both shoulders.

She noticed marked symptomatic improvement with treatment and was so enthusiastic that she invited me (with the help of her rheumatologist) to address the New York City chapter of the Fibromyalgia Society support group. Although many of the attendees expressed interest in the TMS program and its ideas, only a handful ultimately made an appointment to see me in the office.

A few months later, Ms. G called because she developed left lateral heel pain. Her mother, who had been ill, had recently passed away. She rapidly saw the mindbody connection, and the pain resolved.

Another patient, Ms. M, was a thirty-five-year-old attorney who had pain in the buttocks, low back, thighs, and legs. She also had the fibromyalgia diagnosis, even though a lumbar MRI revealed multiple herniated discs. Wisely, a neurologist who knows something about TMS referred her to the program and discounted the herniated discs as pain generators. The patient was suspicious of the validity of a structural diagnosis, as many are when they realize that the pain often changes location. She had all but eliminated any athletic activity due to the symptoms. She was sad and angry because she loved to play tennis.

Her history includes mitral valve prolapse, migraines, and hay fever. She acknowledged being a perfectionist and people pleaser. She felt an enormous amount of job stress and hated the frequent traveling associated with her job.

Her examination was remarkable only for tenderness at thirteen of the eighteen fibromyalgia tender points. She embraced the TMS diagnosis and made a rapid symptomatic improvement.

MR. R'S STORY

Mr. R, a recovering alcoholic in his early forties, came to see me after "injuring" his lower back while performing his job as a stagehand for a major television broadcasting company. As many TMS patients do initially, he immediately concluded that the pain was a direct result of lifting heavy stage objects. In fact, he filed a workers' compensation claim. He proceeded to do what most pain patients do, that is, seek orthopedic attention and obtain a spinal MRI that revealed two small lower lumbar herniated discs. Not long after embarking on a course of physical therapy that proved fruitless, he began to discuss his plight with some friends, including those in recovery. A couple of them cited their successful therapeutic experiences with me, and he decided to contact me for an appointment.

Actually, he was one of the few TMS patients who came to the office without my knowing he was a TMS patient. In fact, when I saw "workers' compensation" as the insurance, I confess saying a small "ugh" to myself, as I expected him to be the type of workers' compensation patient who would be structurally focused, possibly malingering, and resistant to the concept of getting better (not to mention the concept of TMS).

As it turned out, he was wonderful. He was intelligent and well spoken, and had good insight to the mind-body connection. At some point during the consultation, I had to tell him that I would be uncomfortable seeing him under the guise of workers' compensation because I thought he had TMS, not a work-related physical injury. My submitted report to the workers' compensation board would not be helpful to his claim. Accepting of the diagnosis (and a strong people pleaser, to boot), he agreed with the diagnosis and the decision to work outside the workers' compensation parameters.

He and other twelve-steppers have embraced the TMS concept readily and decisively. While Dr. Sarno and I consistently remind patients of the logic behind the diagnosis, twelve-steppers seem to have made a connection between "having faith" in the validity of the diagnosis and giving themselves up to a higher power. Other attractive facets of the TMS approach include the relative nonreliance on painkillers but, perhaps more important, the connection between "anger swallowing" in the process leading up to substance abuse and the repressed anger leading to mind-body symptoms.

Mr. R, like other recovering alcoholics, had a tumultuous childhood including an alcoholic/rageaholic father who physically and emotionally abused him and his family. The father was quite successful at instilling a profoundly low level of self-esteem in the patient at an early age. He developed a hypertrophic superego at an early age and acknowledged being quite the perfectionist and people pleaser. He described himself as someone who apologizes to the person who steps on his foot. The patient became angry and sad at times during the second lecture on psychology and treatment because the lecture material speaks to the early childhood trauma of parental commission and omission that can set the stage for mindbody disorders including TMS. He also suffered from asthma since childhood.

He realized that his pain really began when a series of stressful life events occurred within a short time period. His mother (who he described as an angel) died, his wife left him, and a sibling contracted a serious illness. Before the "injury," he had recently been promoted to a job that "people would kill for." Despite all this, he became confident in the diagnosis and resolute in the TMS treatment approach and made a rather prompt recovery.

OBSERVATIONS AND ADAPTATIONS IN MY TREATMENT PROTOCOL

Lectures are an important part of the education program that is the cornerstone of TMS treatment. This is described in chapter 4. The number of lectures has varied over the year. In recent years, Dr. Sarno and I have used two lectures: the first on the physical aspects of TMS and the second on the psychology and treatment. Dr. Sarno delivers them both in one evening. When I tried that, some of my patients fell asleep during the second hour. After waking some of them up, I asked my audience if I was presenting too much information at one time. Most of them, including my non-Medicare patients, said yes. Since then, I lecture to my patients twice, the anatomy/physiology/medical matters first, followed next time, usually within several days, by the psychological and treatment aspects. Anecdotally, my patients have fared much better with this approach. As a noncontrolled study, I invited some of my last "two-lectures-in-one-night" patients to attend them subsequently and separately, and they were much happier and felt that they absorbed the information better. I do not want to sound "age-ist" in any way; however, my Medicare population seems much happier when I lecture to them in two parts. Here is an excerpt of a letter from one Medicare patient who liked this approach:

> I attended your lectures last summer [note: she wrote this one year after treatment] and took copious notes that I have read and re-read. I identify with one thought after another. The book and the notes fit me like a glove. My sciatica is gone and I am back to swimming, golfing, and more (and I am eighty-nine years of age!).

While Dr. Sarno lectures in a group setting, I decided some years ago to lecture to most of my patients individually. I did this because I wanted to spend my nights with my family, and I had some difficulty reserving rooms and projectors in the building. In rare cases, I have lectured to a handful of patients at a time if they express an interest in a group setting. Dr. Sarno taught me many years ago that there is something therapeutic for a TMS patient attending a lecture with other TMS patients, including the patients' not feeling isolated and being able to share their thoughts, feelings, and experiences with one another.

MY EXPERIENCE WITH THE ORTHODOX JEWISH COMMUNITY

Dr. Sarno and I have helped patients from the orthodox Jewish communities of New York City. I have worked mainly with rabbinical students. They are energetic and enthusiastic about TMS and often are referred to me by senior rabbis in their communities. They relate easily to the notion that their chronic neck/back pain is a reaction to their extreme perfectionism and not the long hours studying at a desk.

THE PAIN OF SAYING NO (ESPECIALLY IF YOU'RE A PEOPLE PLEASER)

One of the most frustrating aspects of practicing TMS medicine is declining to treat a patient who might have TMS or an equivalent but I'm just not sure. This is a rare occurrence due to the phone screening process and the careful history taking in the office.

One patient was a young man in his twenties who presented with hand pain. I thought over the phone that he might have carpal tunnel syndrome or trigger finger, two diagnoses that I've treated successfully with the TMS method. However, in the office, he pointed to his swollen fingers that looked like sausages. In fact, he had the classic "sausage digit" appearance of psoriatic arthritis that was confirmed by a rheumatologist. The patient was quite frustrated because he had similar personality characteristics to the typical TMS patient and he so much wanted to believe that the TMS treatment approach would prove effective. Perhaps it ultimately did.

THE MARTY GLICKMAN STORY

Mr. Glickman was a seventy-six-year-old gentleman who presented with intermittent bouts of lower back pain that began at the age of eighteen when he attempted to make the 1936 U.S. Olympic track and field team. The bouts came about every four years and lasted less than one week. He never underwent spinal surgery. A lumbar MRI performed soon before the appointment revealed severe spinal stenosis. Standing and walking were painful and limited. He acknowledged being a perfectionist and people pleaser. He had recently retired from sports broadcasting—he had been a long-time radio presence in New York as the voice of the Knicks, Giants, and Jets beginning in the 1940s—and felt as if he were getting old. His neurologic examination was normal, yet he had marked bilateral lumbar and buttock tenderness. He began to feel very well after the consultation and treatment lectures, yet he had intermittent setbacks. A few months later, a phone conversation and a *New York Times* article explained why.

Despite being one of the premier sprinters in the world, Mr. Glickman had been scratched from the 4 × 100-meter relay event because he was Jewish and the desire by Avery Brundage (chairman of the U.S. Olympic Committee, whom Glickman claimed was anti-semitic) not to offend Adolf Hitler. Fifty years later, he went to Berlin for the golden anniversary celebration of Jesse Owens's four gold medals. In Marty's words:

> I went down into the well of the stadium and walked along the backstretch of the track, the portion I should have run so many years before. I stopped and looked across to the far side where Hilter and his entourage had watched the games. Fifty feet to the right was the section reserved for the athletes where I had watched the races. Suddenly a wave of rage overwhelmed me. I thought I was going to pass out. I began to scream every dirty curse word, every obscenity I knew.
>
> "How could you no-good, dirty, so-and-sos do this to an eighteen-year-old kid, to any young man who worked so hard to get there, you rotten SOBs." Slowly, after two or three minutes, I began to calm down.
>
> For forty-nine years, that anger and frustration, that rage had

been inside me. Being there, visualizing and reliving those moments, caused the eruption which had been gnawing at me for so long and which I thought I had expunged years ago.

It was the only time I'd been back to Berlin's Olympic Stadium. This Saturday I'll be back there again. I wonder what my reaction will be.

That time, he actually sat in Hitler's box in Olympic Stadium and felt some additional anger. His symptoms worsened temporarily in Berlin and upon his return to New York. He started to feel better over time.

CONCLUSION

My story about becoming a TMS physician is not unlike the Zen expression stating that when the study is ready, the teacher appears (Dr. Sarno). One huge advantage of training at the Rusk Institute was that I had the next best thing to reading Dr. Sarno's book—Dr. Sarno himself.

There is much to be learned about the basic science of TMS. Use of PET scans and/or functional MRI scans may elucidate some of the specific brain changes in the pre- and post-treatment patient population. Noninvasive oxygen tensiometry can document improvements in muscle oxygenation in symptomatic regions pre- and post-treatment. However, the physiology of TMS is relatively unimportant. It is basically a psychological disorder with physical symptoms. Undoubtedly, the brain could create pain in a variety of ways, so focusing on the physiology is unnecessarily diverting from the important question: why does the brain-mind do it? There is no doubt we have not written the last word on that question. To quote the Cy Coleman and Carolyn Leigh song, "The Best Is Yet to Come."

SEVEN

A RHEUMATOLOGIST'S EXPERIENCE WITH PSYCHOSOMATIC DISORDERS

Andrea Leonard-Segal, M.D.

Andrea Leonard-Segal, M.D., graduated from George Washington University Medical School, where she was inducted into Alpha Omega Alpha, the medical school equivalent of Phi Beta Kappa. She is a board-certified internist and rheumatologist and is Assistant Clinical Professor of Medicine at George Washington University Medical School. Currently, she helps people overcome chronic pain at the Center for Integrative Medicine at George Washington University Medical Center.

We are extremely fortunate to have Dr. Segal's contribution to this volume. She is an erudite rheumatologist and a clinician of great warmth.

At the time I attended medical school women physicians were a rarity and I took my training very seriously. Though I chose to specialize in internal medicine and then to subspecialize in rheumatology, I had been attracted to psychiatry as well and had done very well in that training in medical school. I chose rheumatology in part because it offered the opportunity to know patients over a period of time. Rheumatologists treat people with a variety of arthritis and muscle problems; pain in various locations, including the back and neck; and physical conditions that are the result of a noninfectious inflammatory process. Even though the causes of many conditions that we treat have yet to be understood, and many are still not curable, rheumatologists traditionally have helped people to feel better and function better. I have enjoyed conversing and communicating with my patients. I continue to learn so much from taking care of patients and to marvel at the coping skills of those who have to face difficult physical challenges, like people with severe rheumatoid arthritis.

In addition to patient care, my career in medicine has encompassed clinical research and the teaching and training of medical students, medical residents, and rheumatology fellows in the setting of an academic teaching hospital. Having the responsibility of teaching someone else is the best way to learn because the job demands that one master the material completely and crystallize concepts so as to explain them to someone else. While teaching physicians in training, I found that I was often bumping up against the need to explain medical causes and therapies for numerous common disorders. Many of these explanations did not make sense to me, even though they were "standard care." This was not a comfortable situation. With further investigation, it was clear that the accepted, "gold standard," diagnoses and treatments were in fact not based on sound science. In fact, often the logic on which these therapies were based was flawed (and this has not changed over the last fifteen years). It was not surprising that patients often did not respond well. But what to do to improve this situation was not apparent.

As a physician, I have been privileged to take care of thousands of patients. For me, there has been nothing more rewarding than helping patients become pain free so they could live normal, comfortable lives. I am commonly rewarded now when I treat patients who have back and neck pain, tendon and ligament complaints, and other painful disorders not characterized by observable inflammatory changes. However, there was a time during my earlier years as a physician when I did not enjoy taking care of patients who were living with chronic pain because I could never predict whether I could help them. They were in misery, living with various degrees of frustration, loss of control, and feelings of hopelessness. Often, my interventions, which were "standard care," did not succeed in pulling them out of these doldrums because the diagnoses and treatments for many of the pain conditions were nonspecific and did not promise a successful outcome. I was bothered that these standard care treatments did not make logical sense to me. I could not understand, for example, why one would prescribe physical therapy for a patient with bone spurs. Ultrasound and massage certainly would not make this condition go away. Why would people improve? It confused me that I would see an x-ray of an abdomen of a patient and the spine would show lots of arthritic and disc changes, yet the person had never had a backache. It confused me that I would see people who had essentially normal radiographic evaluations and normal physical exams but excruciating knee pain. On the other hand, another patient with substantial degenerative changes of the knee, had very mild symptoms, perhaps just a little stiffness.

I did not understand why we would make a commonly accepted diagnosis of a pulled muscle in the back. The diagnosis was completely presumptive; there were no physical changes to document, no signs of injury, no bizarre physical activity that preceded the onset of the pain. It did not make sense. Lots of things the medical community was doing with regard to this type of medical condition did not make sense to me.

Even when people improved, it was often temporary. I remember treating one man seventeen years ago who was in his thirties and suffered with low back pain. He improved only if he stopped doing things. The minute he resumed work or any type of physical activity the pain would recur. He had tried a variety of noninvasive treatments, and the surgeons did not consider him to be a surgical candidate. He became increasingly depressed and spent lots of time lying down, and I was running out of options to offer him.

At the same time that I was taking care of him, I, myself, developed a backache for the one and only time in my life. It dragged on for a few months and I could not figure out the reason for it. I saw a few physicians who offered standard diagnoses and prescriptions that did not make sense and were of no value. The continuing pain eventually led me to John Sarno, M.D., at the Rusk Institute, New York University Medical Center, first as a patient and then as a colleague; thus began the most profound and rewarding medical education that I had experienced both personally and professionally. This education became a new pathway to benefit patients in a simple, noninvasive, and confidence-building way.

So, for the last fifteen years, I have been fortunate enough to experience the professional satisfaction of moving people—even those who are the most severely ill—from a state of debilitating pain and hopelessness to a pain-free, completely functional life. The frustration I experienced taking care of the young man with the recurrent back problem is substantially a thing of the past. It is gratifying to see people who have been dogged by intermittent attacks of pain break the cycle. It is a pleasure to watch patients who thought they would forever be drug dependent to function, be able to shut the door to their medicine cabinet. It is a pleasure to help people who have had to stop working return to work. It is a pleasure to help people who have suffered from low self-esteem and who have put too much pressure on themselves to succeed in everything, every day, gain confidence and learn to become more satisfied with whom they are. Over the past few years, I have come to devote my entire private practice to these types of patients.

Patients for whom we care are our best teachers; we cannot fool them. When they do not improve, they are teaching us either that our treatments are ineffective or that we are making an incorrect diagnosis. We have made many advances in medicine. We have developed cures and very effective treatments for many complicated and life-threatening conditions. Why, then, have we done so poorly curing and treating common, non-life-threatening problems such as back pain, neck pain, irritable bowel syndrome, and tendon and ligament pain syndromes, for example? Our failures should tell us that we need to think "outside the box" about these conditions. We need to review what is common to patients with these conditions and assess what we may be missing.

WHAT DO PATIENTS WITH THESE CONDITIONS HAVE IN COMMON?

Their symptoms are often triggered by nothing or a very minor physical incident that could be construed to be normal physical activity. Even though the patient may think he has injured himself, there is no objective evidence to support this contention. For example, the involved area is not bruised, bleeding, swollen, red or exceptionally warm to the touch. The physical complaint does not resolve in a timely way like a true injury; the pain lingers unlike what would occur with a true injury. Further, the patient has no physiologic condition that would interfere with the ability to heal normally and reports that his cuts and bruises heal just fine. A broken bone, a ruptured tendon, or a postoperative site, for example, would heal within a couple of months. That is what our bodies are designed to do; they are quite resilient. But often, a patient who has experienced a true injury that has healed normally continues to have pain and does not think about the lack of logic of the continuing pain until it is pointed out to him.

Despite how awful they feel, patients usually have normal physical examinations, except for tender areas in low back and buttock muscles and restricted movement in the back, neck, or certain joints in their arm or leg due to the pain. Patients will often note that those tender spots improve with massage and that after stretching their mobility improves.

Blood tests are normal in that they do not show any evidence of inflammation or chronic disease. They do not suggest a serious, life-threatening diagnosis. The changes on x-rays, scans, and other tests often poorly predict the symptoms. Often, the symptoms are very severe and the diagnostic tests are completely normal.

Thus, even though the patients look very healthy from the view of the physician, they feel frail and vulnerable. I often tell my patients that they need to come to view themselves as I see them, healthy and strong, rather than as they have been seeing themselves. Patients are often surprised to hear this comment from me. Generally, I am the last in a long line of doctors they have seen and none has said anything like this to them before.

I have observed that the pain often goes away only to appear in a new location. The physical symptoms seem to jump from one location to another or sometimes they accumulate, one on top of the other, until everything seems to hurt.

The cause of the physical discomfort is mostly of questionable structural cause, though patients feel like something is structurally wrong. The medical literature attributes up to 85 percent of low back pain as being of unclear cause (Deyo, 2002). As a matter of fact, for decades the medical literature has been replete with articles that point to the fact that structure does not predict pain (Magora and Schwartz, 1976, 1978, 1980; Boden et al., 1990; Jensen et al., 1994; Deyo, 1994). However, the conventional interpretation of these studies is to contort them and attribute the pain to structural abnormalities.

These patients also share certain physiologic and psychobehavioral aspects. They often experience physical events that they think triggered their pain episodes. When explored, there is frequently a relationship between stressful life events or experiences and the onset of physical symptoms. Many patients describe a sense of tension and the fact that, in general, there are not too many aspects of their lives that are "going well." Even though there is no life-threatening problem or objective measure of illness, people think of themselves as fragile or unwell. They are obsessed with their symptoms, often aware of their pain or their body at some level 100 percent of the day. They are terribly fearful. They are very fearful of "injury" and that they will be permanently disabled. They are afraid to engage in many normal physical activities, even during periods when the pain may have abated. They often think that they are easily injured. Fear drives the way they do or do not engage in physical activities. Some patients are so afraid that they essentially stop doing everything physical and are consciously aware of virtually every physical motion they make and how their body parts are aligned with respect to one another. They feel out of control because they expect the pain to occur as a consequence of what they do or do not do. By contrast, this degree of fear and obsession with physical symptoms is not typical, in my experience, even among patients with serious, deforming arthritic conditions like gout or rheumatoid arthritis.

HOW SHOULD THE MEDICAL COMMUNITY DETERMINE WHETHER A TREATMENT IS EFFECTIVE?

The majority of the conventional treatments for the physical complaints described here have not been verified by high-quality, double-blind, randomized, placebo-controlled, prospective clinical trials, but they should be. This is called *evidence-based medicine*. What does this mean? In medicine, we try to determine whether a drug or a procedure works by comparing it to a placebo control or, perhaps, another treatment that has been proven to work, an active control. A placebo treatment can be as a comparison of effectiveness, or a new treatment could be measured for effectiveness against a standard one. This necessitates that the study in which the treatment is tested be designed in a way that will not influence the results and that enough people participate in the study so the results can be meaningful. The study should be conducted ethically and with the informed consent of study participants. Ideally, neither the patient nor the study investigator should know the treatment the patient is taking during the study so that expectations do not bias the study results. This is called *double-blinding* the study. To accomplish this may mean that the investigator cannot be the person who is actually administering the treatment. The patients should be allocated to a particular treatment group by an uncompromised numerical scheme, randomized as to treatment. Clearly defined objectives should be tested. The patients should all be entered into the study under the same study conditions and watched carefully from that point forward. This evidence-based approach should be relatively easily accomplished for the majority of physical treatment approaches (e.g., physical therapy, injections, oral medications, acupuncture, and surgery).

Sometimes one well-conducted study may provide convincing information, but it is always best when more than one study provides confirmatory information. It is better when a study is conducted by multiple investigators in multiple medical centers so that enough people can be enrolled to provide statistically and clinically meaningful information. Sometimes the number of people enrolled in a study is small, but the study has redeeming features. If there are other studies looking at the same medical issue and conducted using the same methods, the data from these studies can be statistically pooled by a technique called *meta-analysis* to provide useful clinical information.

Ironically, the treatments for many of the pain conditions described in this book, although accepted as "the things to do," have not passed these evidence-based tests of scientific rigor but, even so, are still prescribed. Sometimes the studies evaluating whether treatments really work are often flawed because investigators conduct them in ways that bias the results, or the number of patients who enroll and participate is very small. So we need to ask ourselves, "Why do we treat the way we treat?" It is difficult to figure out the answer to that question if we explore the literature. Although it is not possible to do this for each condition we are considering, let us take a look through the medical literature and get a sense of what it has to say respecting this great public health problem.

WHAT DOES AN OVERVIEW OF THE MEDICAL LITERATURE TELL US ABOUT TREATMENTS FOR LOW BACK PAIN?

For the past fifteen or twenty years a research method known as meta-analysis has been employed to survey the published medical literature on a given subject. The following reviews followed that method in looking at a variety of treatments for low back pain.

One group of investigators (Furlan et al., 2001) concluded that the quality of the individual articles in the medical reviews on chronic low back pain varied considerably and that there was a need for high-quality clinical trials to determine the value of different treatments.

Van Tulder and colleagues (2000) reviewed thirty-nine studies on the value of exercise treatment and found that it was hard to draw firm conclusions because the studies were so different from each other. However, they said exercise treatment for acute low back pain appeared to be no better than no treatment or other active treatments. For chronic back pain, there was mild evidence that exercise could help people get back to work and be more active but no conclusions regarding pain.

Jellema and colleagues (2001) looked at the value of lumbar supports and concluded that properly conducted studies need to be done in order to determine whether they were helpful.

Transcutaneous electrical nerve stimulation (TENS) has been widely used to treat chronic back pain. Data from five studies were pooled, and the analysis found no evidence to support the use or nonuse of TENS (Brosseau, 2002).

What about remaining physically active when there is acute back pain? A review of four studies (Hagan et al., 2002) decided that it didn't make any difference but that, because staying in bed could be harmful, people with low back pain and "sciatica" should be advised to stay active.

When the question of whether to do surgery for "degenerative" (normal aging) changes in the lumbar spine was assessed, Gibson and colleagues (2000) found that sixteen published papers did not look at pain relief, disability, or capacity to return to work, but focused instead on technical surgical matters. They concluded that there was no evidence that any form of surgical decompression or fusion was helpful when compared with no treatment, a placebo or nonsurgical treatment.

In an article four years later, Deyo and colleagues (2004) reviewed the literature and stated that spinal fusion is effective for certain rare conditions, but its usefulness for common conditions, such as degenerative disc changes, is unclear. They recommended that research efforts should shift from examining how to technically perform the operation to examining which patients can actually benefit from it. Surgical complications are common.

In 2003, Assendelft and colleagues reviewed the published literature to assess the value of spinal manipulative surgery for low back pain. The authors learned that for acute or chronic low back pain, spinal manipulation therapy had no advantage over general practitioner care, pain relievers, physical therapy, exercises, or "back school" (how to care for your back).

Van Tulder and colleagues (2001) reviewed the value of psychological behavioral treatment for the treatment of chronic low back pain. This is a kind of psychotherapy in which patients are helped to deal with the day-to-day reality of the pain and how to cope with it. They found six good studies and concluded that behavioral therapy was effective, but they could not identify the specific patients that would benefit most from the treatment.

It is disturbing to find that despite the dogma that surrounds the treatment of low back pain, there is little science that supports any particular approach to treatment.

WHAT DOES AN OVERVIEW OF THE MEDICAL LITERATURE TELL US ABOUT TREATMENTS FOR CARPAL TUNNEL SYNDROME?

Since the middle of the 1980s this disorder has grown to epidemic proportions. When fourteen studies were reviewed and reported in the *Journal of Neurology*, the conclusion was that it was impossible to recommend any particular type of treatment and that more high-quality studies are needed (Gerritsen et al., 2002).

In recent years, there has been a steady increase in pain in places that were never the clinical problem that they represent now. I have found papers that deal with heel pain, the use of orthotic devices for certain kinds of knee pain, and the treatment of shoulder pain and Achilles tendon tendinitis. In a total of fifty-six papers dealing with these medical conditions, no clear-cut recommendations for treatment could be made in any of them.

WHAT DOES AN OVERVIEW OF THE MEDICAL LITERATURE TELL US ABOUT TREATMENTS FOR FIBROMYALGIA?

A meta-analysis assessed sixteen randomized clinical trials that looked at exercise as treatment for fibromyalgia (Busch et al., 2002). It was concluded that supervised aerobic training improves physical capacity and fibromyalgia symptoms. This can be interpreted to mean that fibromyalgia patients should not give in to their aches and pains and should keep moving.

WHAT DOES AN OVERVIEW OF THE MEDICAL LITERATURE TELL US ABOUT ULTRASOUND AS A TREATMENT FOR MUSCULOSKELETAL DISORDERS?

Among thirty-eight studies addressing this question, Van der Windt et al. (1999) found that thirteen were adequately conducted clinical trials. There was little evidence to support the use of ultrasound treatment of musculoskeletal disorders. Conditions evaluated included tennis elbow, shoulder pain, degenerative arthritis, and myofascial pain, among others.

WHAT CAN WE LEARN FROM THE ABOVE LITERATURE REVIEW?

Put bluntly, many of the structurally based treatments that are commonly used for these pain conditions do not work. Others have not passed the test of scientific rigor, so we do not know whether they work or not. It is a disconcerting situation, to say the least, and explains why we have so many patients who suffer repeatedly and for long periods of time. Why should structurally based treatments be ineffective if the cause of these pain conditions is structural? Why should it be so difficult to obtain good clinical information about these treatments? It is not difficult to design a clinical trial to compare a medication or other treatment modality to a placebo or to determine if it works to treat pain or cures the underlying cause.

ADDITIONAL OBSERVATIONS ABOUT THE MEDICAL LITERATURE

Studies that report relief from pain do not often tell anything about the patients' willingness to engage in physical activity. They do not determine if the patient is still afraid and feeling vulnerable to injury. Follow-up studies heretofore have been misleading because they have not looked at whether the person develops pain in a new area once the old pain is "cured." Because these elements of fear and moving pain are characteristic of the patients with these pain syndromes, investigators should design studies to see if these symptoms improve as well.

What if the cause of the pain is not structural? What if the pain is initiated by psychological factors? This is an unconventional notion and would explain why structurally based treatments would not be curative. It certainly makes sense to consider psychological causation as the basis for the symptoms and signs discussed in this chapter. Furthermore, structural characteristics are rigid and fixed. For example, a bone spur does not move, so why does the pain move to a new place from the site of a bone spur? The severity of the pain is way out of proportion to the magnitude of the incident to which it is being attributed. The sense of fear is disproportionate with the severity of the medical condition, yet people often feel hopeless. The obsession with the pain is distracting and makes it difficult for the patient to concentrate on work.

HOW CAN WE STUDY A PSYCHOLOGICALLY BASED SYNDROME?

Now it is necessary to tackle a dilemma. As discussed earlier, the medical literature is not very good or very informative with regard to diagnosing or treating the pain conditions under discussion, despite the conventional medical wisdom. Thus, our own diagnosis of these same conditions—that they are psychogenic and should be treated according to the precepts of mindbody medicine—is not up against very worthy "opponents." But can we use classical prospective, placebo-controlled, randomized, blinded methodology to study a psychogenic problem? How can we apply the standards of evidence-based medicine in this situation? Is it possible to do better in studying this approach than has been done? It is difficult because psychological treatments do not easily lend themselves to the ideal clinical trial methodology. How can we conduct studies to see if psychological approaches can cure this condition? Patients with TMS must be psychologically open to the diagnosis to improve. They must be ready to renounce the idea that their cure is to be found in structural or chemical means. Thus, it would be exceedingly difficult, if not impossible, to conduct a study in which patients with the same condition are randomly assigned to different treatments, one of which is the TMS treatment. Because getting better depends on accepting the TMS diagnosis, most patients assigned to TMS treatment would not improve because they would not be able to accept the diagnosis.

However, in the case of TMS patients, they can act as their own controls, and they do. If a patient has suffered for a very long time, has failed with other treatments, and gets better with an educational program that teaches him/her about the physical and psychological dimensions of TMS, that's success. Success is measurable by visual analogue scales and other validated measures of pain and function. If the pain dissipates and if the patient engages in all physical activity heretofore shunned out of fear and does not develop recurrent pain, this is success. If the pain moves from the original location to another during the course of treatment, this is another confirmation of the diagnosis. If the patient regains control of his physical body and is no longer afraid, his TMS is cured.

WHAT IS IT LIKE TREATING PATIENTS WITH TMS?

As a board-certified rheumatologist, trained in the classical approach to treating patients with musculoskeletal pain syndromes, when I first began to make this diagnosis of emotionally induced pain, I was very much aware of entering new territory. Among the first patients that I diagnosed and treated as an emotionally based pain patient was a man in his early twenties whose low back pain had not improved after surgery on a herniated disc. He appeared tense and described that his life was filled with complications at the time; there were many important areas in which he felt out of control. He had read one of Dr. Sarno's books before seeking me out and had "seen himself" among the pages. He was intelligent and had diagnosed himself. All he needed from me was confirmation. By the time I saw him for a follow-up visit two weeks later, he had overcome his pain and was more physically active than he had been in months. He was relaxed and smiling and within a few months took a hiking trip lugging backpacks. This was a patient who had not needed surgery.

I will describe some of the other patients I have seen. One middle-aged man (married with children) had to be driven to his office because he could not sit and had to do paperwork reclining on a sofa. Over the preceding three years he had suffered first from neck pain, a frozen shoulder, and then low back pain. He became more and more anxious, had panic attacks, and could not sleep. Eventually, he became housebound. With the neck pain began a saga that led to many doctors, a chiropractor, an acupuncturist, and an osteopath. He took many different medications and injections and had physical therapy. Fortunately for him, his MRIs and CT scans were essentially unremarkable, but this did not stop one surgeon (from a highly regarded institution) from recommending surgery to cut a nerve he thought may have been the culprit responsible for the pain. The patient wisely chose not to have this procedure. No one was able to give him a clear reason for the pain, and the people he saw approached his symptoms as if each was unrelated to the other. No one addressed what may have been happening in his life. Because of the anxiety and difficulty coping, he began seeing a psychologist (a good idea) and began to become aware of tremendous rage he had harbored toward his wife and his parents. By the time he saw me, the diagnosis of emotionally caused pain made sense to him and he began to get better. At the two-week follow-up visit, he was able to sit up in the office instead of lying down and had become significantly more physically active. Within a month he was able to drive again. His last visit was seven weeks after the first. He was moving well along the path to becoming pain free. He continued to see a psychologist. He contacted me from time to time to update me on his life and express his happiness at how nicely things were going.

A very interesting patient and tender human being whom I treated was a woman who developed excruciating pain in both of her feet after a walk. She had been suffering for several years and had seen orthopedists, rheumatologists, neurologists, and a vascular surgeon, some from very highly regarded institutions. She had received many diagnoses, ranging from neurologic problems to trouble with a bone in the foot. She had lots of diagnostic tests. Her pain had intensified over time such that she was living in a wheelchair. Her feet were the first thing she thought about in the morning and she was aware of them all day long. She was afraid to walk or to do anything that put pressure on her feet. She could not drive because of pain and was taking ten different medications and dietary supplements to alleviate the symptoms, but they obviously were not working. She had come after reading one of Dr. Sarno's earlier books and saw herself in the book, in part because she had suffered from other types of pain in the past. She had seen a psychoanalyst in the past and so was introspective to some degree. Her personality fit the TMS patient to a "T." There were many aspects of her life that were not going well, socially and professionally, and the onset of the foot pain was temporarily related to a particularly difficult time emotionally.

On physical examination of the feet, there was nothing abnormal to see. The diagnosis was clearly emotionally based pain and, to the patient's great credit, she accepted this, was relieved to hear it, and was able to work with it. When she returned for her next visit in a few weeks, she was out of the wheelchair, was driving her car, and had resumed short walks. The pain was receding. We were able to taper and discontinue all medications. Over the next month it essentially resolved completely. It is sad to consider how long this patient suffered while going from one doctor to the next until finding the answer. But it was wonderful to see her life resume in such a pleasant way.

It is important to note that there are many other physical disorders that serve the same psychological purpose as TMS. They are called equivalents of TMS. A patient whom I had successfully treated for low back pain returned with the complaint of frequent bowel movements. She had a new job, some of the duties of which made her very nervous. She had been diagnosed with irritable bowel syndrome and noted that it was difficult to drive all the way to work without stopping at a restroom. As with the low back pain, she had become obsessed and fearful of the abdominal symptoms. She was afraid that she would not make it through staff meetings without having to run to the bathroom. She agonized about how this might affect her advancement in the company. Since she was sophisticated with the emotionally based pain concept, I did not have to work too hard to convince her that the abdominal problems were of a similar nature. We focused on overcoming the fear that had developed with regard to her bathroom needs. She recognized that they were a distraction for having to perform publicly at work and realized how self-esteem issues were intertwined with the bowel problem. She began to challenge the symptoms, and when she thought she needed a bathroom, she would delay finding one and would focus on things other than her belly. Often, the urgency would pass. Over the weeks, she was able to cure her symptoms, gave a successful presentation at a meeting, and felt good about herself.

In the same vein, there was a middle-aged man who had been suffering from severe constipation for several years. This was the opposite clinical manifestation of irritable bowel syndrome. He had seen many doctors and took many medications on a daily basis to be able to defecate. At times, the constipation had been so bad that he went to hospital emergency rooms. One physician had recommended surgery to remove the colon if things did not improve. He was referred to me, and we began to focus on his fear of not being able to defecate and to address the other issues in his life, which included complicated family relationships and job insecurity. He began to see a psychotherapist and began to recognize his extreme perfectionism, the low self-esteem that was behind it, and how the tension in his life related to the symptoms. Over several months, we were able to decrease his obsessiveness about bowel function, improve his bowel habits, and reduce laxative requirements. The last time I saw him, he was off the majority of his medications.

Another interesting person I saw was a gentleman with dry eyes. The symptom had been there for a long time, and ophthalmologists had not found an underlying medical condition to explain it. The patient was so uncomfortable, though, that in an effort to relieve the symptoms one ophthalmologist tied off the return tear ducts so tears would linger in the eyes instead of finding their natural path into the nose. Somehow, he found his way to my office. At first I looked for rheumatic disorders that could cause dry eyes. Finding none, and considering the ophthalmologist's negative findings, this became a very challenging problem—that is, until I realized that he was obsessed with the dry eyes and very fearful of the gritty symptoms and redness that they caused. At one point I asked if he had ever cried at a sad movie. Surprised at my question, he said, "Yes." I asked if tears had fallen down his cheeks, and again he responded affirmatively. If there were an organic reason for the lack of tears, this could not have happened. This recognition was the true beginning of his loss of fear and ability to get better. The eye symptoms began to recede and he started psychotherapy, the true path to healing.

One young man had suffered from recurrent pelvic pain after sexual intercourse that had been evaluated thoroughly by more than one urologist. He carried a diagnosis of chronic prostatitis. Upon listening carefully to his medical history, it was apparent that he had TMS. This information, psychologically challenging his pain, and psychotherapy in combination led to marked improvement so he could enjoy a sex life. Likewise, women with chronic pelvic pain have improved.

I have successfully treated many patients who carry the diagnosis of fibromyalgia. Patients with this diagnosis have more physical symptoms than the patient with only a backache, or neck ache, or insomnia, or irritable bowel syndrome, or ringing in the ears. However, I think of the patient with fibromyalgia as the most seriously affected with TMS and other patients as having less severe TMS. It is almost as though those with fibromyalgia have so many issues from which the pain distracts them, that they accumulate pain in many locations as their defense mechanism. It is important to remember that this is a syndrome, in other words, a set of symptoms. These symptoms, widespread pain, headache, tenderness at specific body locations, fatigue, sleep disturbance, and irritable bowel syndrome, have been linked together by the American College of Rheumatology to define this condition. There are many other physical complaints that people with this diagnosis can have that involve essentially every body system. Other than specific tender points, the physical examination findings will be normal or not particularly specific and there are no characteristic laboratory tests. Patients generally have lots of physical complaints but actually look quite healthy to the clinician. As with other TMS patients, those with fibromyalgia can be helped if they are open to considering and then accepting that there is a psychological cause for their pain. If they are unable to be open in this way, I have not been able to help them.

One very common condition that people with TMS seem to have is carpal tunnel syndrome. This syndrome has become "fashionable" particularly since the advent of computers. However, I have noticed that many patients have it as part of a continuum of replacement pain syndromes, or as one element of lots of different accumulating pains. I took care of a young professional man who had this diagnosis while working at a very stressful and demanding job. He had to stop working and went on disability because he could no longer use his hands to do his job. He was miserable not working, and was getting increasingly depressed about the situation. His personal life had suffered many rocky periods because of a broken marriage and other traumas.

Many of my patients have been children of alcoholics. Many have been children of divorce, and many have been abused. However, the majority have come from families with hardworking, loving parents who conveyed overly high expectations and hopes for their children. These families are not characterized by any particularly unusual dynamics that would stand out in today's society.

Patients with these emotionally induced physical symptoms are often in the throes of decision making. They may find themselves in the uncomfortable place of someone who is uncertain of the best choice to make. The issues holding them back generally have to do with concerns about disappointing someone, or setting standards too high for themselves. Sometimes they are considering a career change. Maybe there is a decision about moving to another town. Perhaps there are fears related to the responsibility of becoming a new parent. They might be considering a divorce or rebounding from one. Sometimes the emotional issues relate to how to communicate sexual orientation issues.

Often, it is difficult for the patient to see his or her way out of a difficult situation. Patients often take responsibility too seriously and have forgotten how to have fun or cannot allow themselves to do so.

There are other emotionally difficult situations that can trigger or perpetuate pain. Maybe there has been a recent death in the family. Maybe the patient has taken on major caretaking responsibilities for a loved one who has become infirm. Maybe a new mother is feeling overwhelmed taking care of her young family and does not know how to find an effective support system. Maybe an adult child has unexpectedly moved home or has dropped out of school. Maybe there are difficult financial concerns. These are just examples of some of the types of circumstances that have tipped my patients over into a pain or other physical symptom syndrome. Sometimes it is difficult to identify a specific cause, but more often than not, a little digging will turn up at least the signs of an emotionally charged association.

It is not unusual for me to have taken care of multiple patients in the same family, husbands and wives, children and parents, brothers and sisters. In one family, I treated pain in the daughter, mother, and father. I treated twins, both during the same afternoon. One had back and leg pain and the other had neck and arm pain. Both had grown up in a household where they felt enormous pressure to succeed, and both had excelled and done just that until life caught up with them and they had to reevaluate where they were headed.

It is important to recognize that TMS is a symptom of life not going well and out-of-control circumstances and emotions. It is not surprising that when one member of a family feels pressures, others do too. The insight that comes from understanding the basis of physical symptoms that stem from emotions, and the insight that comes from recognizing the source of the feelings, can ultimately lead to better family relationships all around.

The patients I see tend to be very hard on themselves. They take responsibility too seriously and forget that, as adults, it is also okay to have fun. They have extraordinary expectations of themselves and of those around them and so are often disappointed or angered by what they perceive to be their own shortcomings or those of others. They always try to do their best, not realizing what a hard job that can be. I often wonder why, in our society, people cannot be "average" anymore. Now in school, a "C" grade is tantamount to failure. Yet, when it really comes down to it, we are all average most of the time and can do our best comfortably only some of the time. Doing our best all the time is exceedingly difficult work and can lead to frustration and buried anger. My patients are often disappointed if others do not notice the hard work they have done. Low self-esteem needs to be fed by compliments from others.

Getting better from TMS is learning how to extract yourself from needing recognition from others and learning how to fill that need yourself. It is about learning to parent yourself in perhaps a kinder, gentler, and more benevolent way than one may have actually experienced during childhood. It is learning to lower your expectations of yourself and others and learning that relationships are easier, more genuine, and form stronger bonds under those circumstances. It is learning that if someone seems upset with you, the feeling is often displaced and you are not at fault. It is learning that it is even okay if someone does get angry with you. It is about learning not to be afraid to take care of yourself psychologically, to say "no" when you want to and "yes" when you want to. It is about learning that almost all of the time that you feel guilty, it is inappropriate, in that you cannot be responsible for taking care of everyone's feelings. People need to learn to do that for themselves.

Patients need to learn that these TMS symptoms are in essence "growing pains." The TMS patient, for example, is the former child who learned that speaking up might aggravate the frazzled adult in his life, who might get angry; this was scary. The TMS patient is the former child who was never praised unless he accomplished something. The TMS patient is the former child who was abused. The TMS patient is the former child who thought that his parents' divorce occurred because he was not perfect. The TMS symptoms come from the psychological conflict that forms in the adult who sees life through the same eyes that he had as a child, when he had little or no power or control over life circumstances, and acts accordingly. The child's view and the behavior that once was a survival technique no longer work.

Sometimes patients are not ready to accept the TMS diagnosis. I have made this diagnosis in patients who have not believed me and have told me that I was "crazy." They have been invested in thinking that their problem was structural or generated by some other physical cause. At some level they were insulted or threatened by this TMS diagnosis. I never try to convince patients who are not open to this diagnosis, because I have learned that when patients erect an emotional barrier, they leave it standing until they are ready to take it down. No one can do this for them. Some of those patients have never returned to see me. Unfortunately, they do not become aware that emotional pain can cause severe physical pain. Others, however, have lived with their symptoms for a while longer, and then one day, for whatever reason, became more introspective about their condition and returned to see me with an open mind. Those patients then were able to get better.

Sometimes when I give this diagnosis to a patient, the patient misunderstands me and thinks that I am saying that the pain is not real, that it is all in the head. This could not be more wrong and I need to clarify things for the patient. TMS symptoms are physical symptoms, but are generated by a feeling. There are many things that can cause back pain. Some examples include infection, cancer, and a fracture from thinning bone. Emotions are another cause.

The mindbody connection has been accepted by some and rejected by others. Unfortunately, physicians are often among the nonaccepters. It strikes me as curious that people will readily connect a feeling of embarrassment with red cheeks or a feeling of nervousness with sweaty hands but are unwilling or unable to make a connection between anger and pain. We do not know what molecules come together to translate a social incident into embarrassment and how that feeling then stimulates the blood vessels in the cheeks to dilate. But we all accept that it happens. Literature has suggested links between chronic stress and health problems. However, in the December 7, 2004, issue of the *Proceedings of the National Academy of Sciences*, Elissa S. Epel and colleagues published an exciting article that strongly affirms that the mind plays a significant role in our physical health. They demonstrated a definitive connection between perceived and chronic stress and changes in telomeres, specific areas of the chromosome that are known to relate to the length of cell life and aging. It is critically important to human health that we shed our stigma about the nature of things emotional. Toughing things out, burying uncomfortable feelings, and living under stress can lead to physical pain and dysfunction and, most important, serious medical illness.

The beauty of the TMS diagnosis is that it is a hopeful one that can result in a true cure. The treatment leads to resumption of full physical activity, the emergence of a more emotionally healthy life, and an education in self-awareness. The patient who has recovered from TMS grows into a happier, more comfortable, more peaceful person who sees new paths toward greater personal fulfillment.

EIGHT

MY PERSPECTIVE ON PSYCHOSOMATIC MEDICINE

James R. Rochelle M.D.

James R. Rochelle, M.D., graduated from Creighton Medical School in 1978 and did a general surgical internship at Southern Illinois School of Medicine, followed by a four-year orthopedic residency at Hamot Medical Center at Erie, Pennsylvania. His training continued with a pediatric orthopedic fellowship at the A.I. duPont Institute in Wilmington, Delaware, completed in 1984. He entered into private practice in North Kansas City; moved to Council Bluffs, Iowa, four years later; and in 2004 moved to his present location at Mena, Arkansas.

To have a full-fledged orthopedic surgeon practicing mindbody medicine and contributing to this volume is as exhilarating as it is unusual. Like Drs. Sopher and Hoffman, Dr. Rochelle has a dual practice: teaching sessions for those who can accept the diagnosis and conventional treatment for those who cannot They all agree, it is challenging.

As a general orthopedic surgeon, I see patients with fractures, sports injuries, workers' compensation and personal injury problems, and many cases of acute and chronic low back and neck pain. I also see patients who need joint replacement surgery. Other common conditions are tendinitis of the knee and shoulder, rotator cuff pathology, chondromalacia of the patella, and carpal tunnel syndrome.

DISILLUSIONMENT WITH "CHRONIC PAIN"

In the early years of my practice I usually referred patients with chronic neck and back pain to chronic pain specialists. The standard treatment was a series of therapeutic injections usually consisting of a combination of local anesthetics and steroids. While sometimes helpful in acute situations, they were rarely useful in patients with pain lasting more than four to six months. Other techniques and a variety of injections were also used.

Pain specialists tend to employ elaborate protocols that require expensive diagnostic tests in an effort to identify specific "pain generators." In their view, chronic pain is the result of "structural abnormalities" and these "pain generators." They reject out of hand a psychosomatic explanation.

When asked about their response to injections, patients often reported some benefit for a month or two and then a recurrence of pain. It was common for patients to return for injections on an as-needed basis, while others came back for weeks. I know of no patient with chronic pain who was permanently relieved of pain by these techniques.

In those early days I occasionally referred patients to multidisciplinary pain clinics. Insurance carriers sometimes authorized this expensive program in the hope that it would end the need for continuing therapy. A four-week program might consist of education, group therapy, psychological counseling, physical therapy, exercise, and so on, administered by a variety of professionals. Though sometimes helpful, the scattered therapeutic efforts confused some patients who were unsure whether to concentrate on exercise, physical therapy modalities, psychotherapy, or cognitive insights. Beneficial effects of the program seemed to wear off over time, with a few showing permanent resolution of pain.

I have found that patients often seek out nonmedical practitioners on their own because of dissatisfaction with conventional medicine. Minor abnormalities of the spine are often treated with manipulation or other physical modalities. The results are mixed. In the light of what I have learned about acute and chronic pain, I believe that improvement in these cases may be due to the passage of time or a placebo reaction. The laying on of hands and a sympathetic, warm interaction between practitioner and patient can also contribute to symptomatic improvement.

Having surveyed the field of "chronic pain," I became convinced that no one had any long-term answers. I no longer wanted to refer patients for treatments that weren't working! I felt strongly that patients were actually being harmed by the attitude of their physicians. Many patients had become convinced that they had a "bad back" or a "bad neck." Doctors would encourage their patients to apply for Social Security disability benefits because they could see no other way of helping them. Some even told their patients that they might end up in a wheelchair. Andrew Weil, M.D., has referred to this negative conditioning as medical pessimism. With such gloomy prognoses, patients became disheartened and lost hope that they could live a normal life again. This negative conditioning is a major reason for the pain epidemic in the Western world today.

Another form of treatment of dubious value is surgery for the back and neck. There is now a large number of surgeons who specialize in such surgery. In addition to surgery to remove herniated disk material, spine fusions are now commonly done, though it has not been demonstrated that they are more effective than the simpler procedure. Indeed, experience treating TMS questions the need for any surgical procedure in most cases. Elaborate surgical hardware has been invented to provide stability for the spine, based on the idea that the back is unstable and, therefore, painful. There is nothing in the medical literature that supports this contention, despite which fusion surgery has increased dramatically in recent years.

One of the consequences of all this surgery is something called the *failed back syndrome*. Despite technically well-informed spinal fusions with impressive x-rays, a large number of the operations end in failure! Most of the literature on failed back syndrome indicates that "psychological factors" are probably responsible for poor results, with failures approaching 30 to 50 percent. The exact nature of these "psychological factors" is poorly understood by spine surgeons, who are nevertheless undeterred in their enthusiasm to perform lumbar spine fusions. They obviously have strong financial and professional incentives to continue doing what they have been trained to do.

PAIN MANAGEMENT

I was becoming increasingly disillusioned with conventional diagnoses and treatments and decided to do "pain management" on my own patients. I knew that my patients had real, physical pain that was interfering with their daily functioning and enjoyment of life. There is a nationwide movement to stop stigmatizing people with chronic pain, and to provide appropriate analgesic medication for those with moderate and severe pain. I welcomed this new emphasis, and educated myself on the use of opiates. Many doctors still reject the idea of providing "narcotics" to people with chronic pain. I reject the pejorative term *narcotic* because it contributes to the stigma attached to the appropriate treatment of chronic pain. I started using opiates more aggressively, encouraging patients to maintain adequate blood levels to control their pain during the day.

Many patients responded favorably to this approach. Their pain scores decreased and their function improved. I encouraged people to lose weight and to walk thirty to forty minutes daily. However, the benefits were only partial. Many people required higher doses of medication. I explained that tolerance to medication is not the same as addiction, and encouraged people to continue with their lifestyle modification. However, the prospect of staying on opiates on a long-term basis was unattractive to me and to many of my patients.

I reviewed the diagnoses of many of my patients and found the usual assortment of spinal conditions: bulging disc, "degenerative disc disease," spinal stenosis, "facet arthropathy," spondylolisthesis, scoliosis, and so on. Seldom was there any significant nerve root impingement. When a "pinched nerve" was suspected as a strong possibility, I looked at the MRI with a radiologist to "compare notes" on the significance of the findings. On occasion, I referred patients to a neurosurgeon for a second opinion. The patients would usually return to state that surgery was not recommended. I was seldom able to attribute the chronic nature and severity of the pain to the reported MRI findings.

Despite a modest improvement with "pain management," I was

at a loss to correctly diagnose and treat my own patients. I was unable to honestly tell people that I could help them get rid of their chronic pain once and for all. Clearly, something was missing. The psychology of chronic pain needed to be addressed.

I referred some people for psychological evaluation. The referral process was a touchy subject because of the inference some people made about my request for an opinion about emotional issues. "Are you saying that this pain is all in my head?" my patients would ask. I explained that the pain was real and physical, but that emotional issues might possibly be part of the picture.

Psychological evaluation and testing often revealed depression and anxiety. Psychiatrists and/or family practitioners prescribed medication for this aspect of the problem and to help people with their sleep. I always assumed the depression and anxiety was secondary to the chronic pain as most of the literature said that it was.

The literature on chronic pain further assumed that most patients were motivated by "secondary gain," that is, the unconscious desire to be cared for, avoid responsibility, or perhaps get money as a result of their condition. I tried to get people back to work as quickly as possible following an injury to avoid rewarding patients for playing the sick role. This rather cynical attitude would often place my patient and me in adversarial roles. Like many other doctors, I felt uncomfortable with this conflict because we sincerely wanted to help our patients get better. Patients were uncomfortable because they perceived that they were being blamed because they still had pain! Quite legitimately, they were not yet ready to return to work because they were still in pain.

I never bought into the idea that secondary gain was the primary motivation for most patients, but I had no better explanation for the "psychology of chronic pain." I had the impression that most everyone with chronic pain had elements of *functional overlay*. This term refers to the various emotional issues that crop up in the context of personal injuries, workers' compensation medicolegal issues, conflict at work, family hassles, and so on. I felt it was normal to experience emotional upset in response to these stressors. I knew that a large majority of patients sincerely wanted to return to work and to resume a normal life as soon as possible. The psychological evaluation had shown most people to be honest and straightforward, with issues of depression and anxiety in some cases.

WHAT'S MISSING?

Obviously, something was missing in my understanding of "the psychology of chronic pain." What was I missing? I conducted an extensive literature search in the field of chronic pain, including the realm of alternative medicine. I read Dr. Andrew Weil's *Natural Health, Natural Medicine.* Dr. Weil strongly endorsed the concept of tension myositis syndrome (TMS). Dr. Weil wrote: "I am a great believer in TMS, having seen a great many cases of chronic back pain disappear as if by magic when people fall in love or otherwise make radical changes in their emotional and mental life." The possibility that chronic back pain is psychosomatic was intriguing to me.

On Dr. Weil's recommendation, I read *The Mindbody Prescription* by Dr. Sarno. The first thing that caught my attention was his discussion on ulcers. I had a vivid memory that ulcers were very common in the 1960s and 1970s. They had strangely gone "out of style" since then, because it was well known that ulcers were stress related. No one wanted to have a psychosomatic condition that could be controlled simply by attending to the issues that generated stress. The understanding that ulcers were probably psychosomatic caused ulcers to become much less popular in the 1980s and 1990s.

I realized from my own practice that I was seeing more and more chronic back and neck pain. There was quite a bit of "tendinitis" or "tendinopathy" of the knee and shoulder that was difficult to explain. Many people had symptoms of carpal tunnel syndrome but were not candidates for surgery. Many people with personal injuries and workers' compensation claims were having long-term pain problems. Dr. Sarno's book explained that these chronic musculoskeletal conditions had now replaced ulcers as the primary stress-related conditions in our modern world. I came to understand that we are in the midst of an epidemic of pain caused by the psychological issues addressed by Dr. Sarno.

The psychological issues had not been adequately addressed by conventional medicine because there was no appreciation of the importance of the unconscious. Dr. Sarno gave primary credit to Freud for emphasizing the importance of the unconscious in all human behavior.

In order to learn more about incorporating TMS theory into my own practice, I visited Dr. Sarno at New York University in January 2002. I sat in on forty-five-minute office visits with five of his new patients on each of two afternoons. I attended the two-hour lecture he presents on a weekly basis. On the second evening, I attended a small group follow-up session for those who were still having pain a month or so after attending a previous lecture.

There were three major eye-openers for me when I compared my own practice to that of Dr. Sarno. They related to the great significance he placed on:

1. The patients' social history

2. The findings of tenderness on physical examination

3. The importance of one diagnosis, one unifying concept: TMS

SOCIAL HISTORY

The most striking difference between my practice and that of Dr Sarno was in the realm of history taking. Dr. Sarno spent a good deal of time obtaining the social history. This includes information about marital status, family history, siblings, and occupational information. Then he asked another question: "How was your childhood?" I had *never* asked that question of any patient in all my years of orthopedic surgery! Some people had good childhood experiences. Some people had experienced significant difficulties. Dr. Sarno allowed the patients to do most of the talking, trying to get a feel for different types of conflicts that people may have had in their childhood. If physical, emotional, or sexual abuse had occurred, we had a brief discussion about its impact. Dr. Sarno tried to get a feel for what kind of relationship people had with their parents or guardians.

The next significant question: "Can you talk about your personality, who you are?" He would zero in on issues of perfectionism and goodism. He developed a sense of whether people are people pleasers and/or peacemakers. He explained that this pressure to be perfect and to be good is enraging to the unconscious mind.

Finally, we discussed issues of life stress that we all have. I was impressed that many people were forthright and insightful about these issues to a large extent. They were willing to accept the idea that stress not only aggravates pain, but that it can actually be the primary source of pain.

As I reflected on my experience in New York, I now realized one of the major reasons I hadn't understood the psychology of chronic pain. I had never asked people about these very important aspects of their lives. Why should I bother to ask people about their childhood? Did it really matter what kind of personality traits my patients had?

Could issues of life stress really be all that important in causing the pain? After all, I was an orthopedic surgeon! I was too busy to take more than fifteen minutes to conduct an office visit for someone with chronic pain.

My previous attitude of neglect and ignorance of the importance of the social history is certainly typical of the vast majority of conventional doctors. I had never learned how to take a good social history. Like most doctors, I usually skipped over this vitally important source of information. Previously, if I attempted to get a social history at all, it was perfunctory and superficial. It covered only items like smoking history, family health problems, and exercise patterns. Inadequate social history taking must be considered a primary reason that most doctors do not appreciate the true psychological factors that cause chronic pain.

Dr. Sarno's books provided a framework for understanding why these aspects of social history are important. He explained to his own patients that there are three sources for unpleasant emotions in the unconscious: childhood anger, emotional pain, and sadness; personality traits such as perfectionism and goodism; and the realities of life. He explained the mechanics of daily study as part of the treatment. People get better by developing a conscious awareness of the sources of unconscious rage and emotional pain.

PHYSICAL EXAM: TENDERNESS

Dr. Sarno's physical examination was instructive. He did a complete musculoskeletal examination, with emphasis on findings of tenderness. I never realized how frequent the finding of tenderness on the outside of the thigh is in patients who have chronic low back pain. In Dr. Sarno's experience, this correlation is approximately 80 percent. Tenderness in the gluteal and lower lumbar region is almost universal in TMS, as is tenderness at the top of both shoulders. Many patients have paraspinal tenderness throughout the lumbar spine, extending sometimes into the thoracic area as well.

The tenderness most people have with chronic low back pain is usually present over a very widespread area. Palpation requires a light to moderate amount of pressure in strategic locations, indicating that the pain is usually in the muscles or the tendons just below the skin. This superficial tenderness indicates that deeper structures, such as the intervertebral disks, are not producing the pain.

Before I knew about TMS, I tended to gloss over palpation of areas of tenderness in performing a physical examination. I usually had a predetermined notion that a person's chronic low back pain was caused by some degenerative condition in their back. Often, I relied on an MRI, assuming that some "structural abnormality" was the cause of the pain. These abnormalities are simply mentioned by the radiologists in their reports because it is their job to report everything they see. It is the job of the treating physician to put all the information together in arriving at a diagnosis. I had been too quick to jump to the conclusion that the reported "abnormalities" on the MRI were the cause of the patient's pain. I was dazzled by the clarity of the MRI technology. Like most physicians, I had a need to demonstrate that the cause of the pain was physical.

I had simply neglected the fact that these degenerative and bulging discs seen on the MRI are normal! We all get them as we get older. Even herniated discs seldom give rise to chronic pain syndromes because the swelling and inflammation from an acute herniation usually settles down within one to two weeks. I had not been doing an adequate examination for tenderness and, therefore, missed out on this most important physical finding. The diffuse tenderness that is present in TMS indicates that the central nervous system, the brain-mind, is involved. This is the only explanation that makes sense. It is anatomically impossible for disk abnormalities to produce tenderness in the iliotibial band, a tendon, and the gluteal, quadratus lumborum, lumbar paraspinal, and upper trapezius muscles, as we see in chronic low back pain caused by TMS.

Tenderness findings are also important in two areas of interest to an orthopedic surgeon: the knee and shoulder. Patients often present with a previous diagnosis of arthritis, chondromalacia patella, torn meniscus, or tendinitis of the knee. Physical examination findings at the knee will often show tenderness of the quadriceps tendon, patellar tendon, and the medial and lateral retinaculum (tendon tissue at the sides of the kneecap). Before I knew about TMS, I had attributed this tenderness to tendonitis or tendinopathy (a chronic form of tendinitis). The exact nature of this tendinitis has never been well explained in the literature. True tendinitis can occur from heavy activity or from athletic activity. It usually subsides within several days to a couple of weeks. Yet many patients with chronic knee pain will have distinct areas of tenderness in the front of the knee from TMS. Since most doctors attribute knee pain to the more well-recognized conditions above, more and more patients are presenting with knee pain that is predominantly TMS. Knowledge that TMS is often the primary source of knee pain allows patients and doctors to avoid expensive, time-consuming, and often risky treatments.

Tenderness findings at the shoulder show a similar pattern of diffuse involvement when TMS is the cause of pain. The upper end of the biceps tendon is a common location for TMS, with tenderness extending several centimeters below the acromion process at the top of the shoulder. Many patients report pain throughout the upper arm, and have several tender areas at multiple locations, especially at the lateral epicondyle of the elbow (the common site for "tennis elbow"). These findings are not consistent with the common diagnoses I used to make, namely, impingement syndrome and tendinitis.

These conditions do occur but typically present with very localized findings or aggravation of pain by the "impingement test."

Rotator cuff tear is another common diagnosis, usually made on MRI. The tear can be either partial or complete in thickness. A complete tear will often produce marked weakness on strength testing due to the anatomic defect. These tears are fairly common beyond age sixty, probably affecting 30 percent or more of all individuals beyond that age. Yet far fewer patients actually present to an orthopedic surgeon for shoulder pain. Many people tolerate the strength deficit rather well and require no treatment. If a patient has significant pain, and the MRI shows a complete rotator cuff tear, other doctors almost invariably blame the pain on the torn rotator cuff.

My experience with TMS leads me to another conclusion. The typical tenderness in a number of different tendons is a sign of TMS and thus a relative contraindication for rotator cuff surgery. If surgery is done, results will be less than optimal, at least for the pain. Again, an understanding of TMS avoids expensive, risky, and unnecessary treatment.

ONE DIAGNOSIS

There is a strong tendency for pain loci in TMS to move from one place to another. I used to see patients with back pain for several months. After resolution of the back pain, I would see them later for shoulder tendinitis and still later for neck pain. I now realize they were having location substitution. This is typical of TMS; Dr. Sarno refers to it as the symptom imperative, meaning that if the psychological need for physical symptoms continues, the brain will continue to produce them until the psychological situation changes for the better.

This is an important revelation for both patient and doctor.

Many people spend a lot of time and money seeing super-specialists for each different area of the body. They often get the impression that they are falling apart physically. They are at the point of near exhaustion from all the various diagnoses and treatments.

Many specialists tend to see the patient only in terms of his herniated disk, his shoulder impingement, or his carpal tunnel syndrome. There is a woeful trend toward "body parts medicine" that fails to see people as individuals. A patient is simply a collection of body parts to many of today's specialists. Patients receive a diagnostic label and receive treatment according to an "evidence-based model." This cookbook approach to treatment may serve the interests of the managed care bureaucracy, but it drives a wedge between patient and doctor. There is very little appreciation of the whole person and no understanding of the mindbody connection. There is little wonder that many people seek alternative treatments for the many different conditions diagnosed by conventional allopathic physicians.

An excellent tool for understanding the whole person is to have the patient complete a pain diagram. I ask them to shade in all the areas of the body where they feel pain or have had pain in the last year. Many people with TMS shade in a fairly large portion of the pain diagram. They have had pain in a number of different locations in the preceding months. I have pain diagrams over several years on many patients. The change that occurs over time makes a compelling case for the tendency for location substitution that occurs in TMS.

Most important, the knowledge that a person has one condition, TMS, and not several different and unrelated diagnoses is a huge benefit. There is no need to spend time and money on multiple diagnostic exams and visits to specialists. There is no need to pursue different chemical or physical treatments.

Patients can be reassured when they have TMS that their back is normal; their neck is normal; their arms and their legs are normal. Reassurance is powerful medicine. People can stop worrying that their next attack of severe neck pain is just around the comer. They will realize that they don't have a permanently "bad back." They can work to reverse the powerful negative conditioning the medical establishment has foisted on them over the years. They can rediscover faith and hope in the future.

It's really a question of diagnosis. That's the heart of the matter, Dr. Sarno has stated. The TMS diagnosis is liberating and empowering for patients. It gives chronic pain sufferers their best opportunity to live a full and rewarding life, free of pain.

IMPLEMENTING TMS IN MY PRACTICE

The approach to each person needs to be individualized to address the issues that are relevant to their particular presentation of TMS. Some people need a lot of discussion about structural abnormalities that may have been diagnosed by other doctors. Many people have significant fear about certain activities that cause pain. Other people have a hard time believing that unpleasant emotions can cause physical symptoms.

I tell people to give the educational program a try. It's natural to have reservations about a concept that is so different from orthodox medical diagnosis and treatment. It's my job to educate them about TMS and to address their individual concerns. Many people agree to adopt an open attitude, and they usually accept my initial invitation to attend the TMS lecture.

My usual procedure is to call the evening before the lecture to remind people of the time and location. I present the lecture once every two weeks. Usually, I have a list of ten to fifteen patients I have seen in the preceding two weeks who have agreed to attend the lecture. Most people state their intention to come when they see me for the initial visit, while a small number state that they are not coming because they reject the concept of TMS. I expect that half or more of my patients will eventually reject the diagnosis. This should not be a source of irritation or hard feelings. I tell people who reject the diagnosis that I respect their opinion, and I wish them good luck, presumably with another physician.

The rate of "no shows" for the lecture averages 30 to 35 percent. I usually give people a reminder call for the next lecture, but if they don't show up a second time, I take this as an indication that they are probably rejecting the diagnosis. Some people who reject the diagnosis are looking for a more conventional treatment. Some people want a "quick fix." Many people want me to prescribe a medication or perform an injection or operation that will cure their pain problem. I can't magically get rid of their pain for them. It is only those who are willing to take responsibility for learning about their condition and doing the daily study program who will be cured of TMS.

As a physician, I have always prided myself on being able to help the large majority of my patients. It is quite frustrating for someone like me to accept that half or more of my patients will reject my diagnosis and treatment. Some patients take offense at the suggestion that their condition is psychosomatic. I avoid using this "P-word" during the initial interview (unless specifically asked about it) because some people believe that psychosomatic means that the pain is "all in my head." I strongly emphasize that their pain is real and physical. The pain is in the body. Despite my best efforts to clarify this point, some people seem afflicted with selective hearing. They hear that the mind is involved in TMS. They seem to believe that I am implying they have some sort of mental illness, and they reject the diagnosis.

CASE STUDY: REJECTING THE DIAGNOSIS

A seventy-two-year-old woman has seen me for left shoulder and arm pain for the past two weeks. On her pain diagram, she shaded in almost all the lateral (outside) aspect of her shoulder, elbow, and forearm. She reported tingling and numbness occasionally throughout most of the arm. She also indicated pain in both buttocks and in her low back, present for several years, but not much of a problem recently.

She had full range of motion and strength in both upper and lower extremities. Motion was also normal in both her neck and low back. Tenderness findings were in the biceps tendons from the shoulder down her arm about four inches from the top of the shoulder. She had tenderness over the lateral epicondyle (tennis elbow area) of the humerus. Several tendons were also tender on the back of her wrist. Tenderness was also present in the gluteal and lumbar paraspinal muscles.

I obtained x-rays of the upper and lower portions of the arm, which were normal.

Her social history was as follows: She had a good childhood, without significant conflict with her parents. She was a perfectionist and a people pleaser. Her husband had died two years ago. She was very close to her sister in Texas. They visited each other every three months, taking turns in making the airplane trip. Her sister had recently gone into a nursing home and was unable to fly to Council Bluffs anymore. My patient continued the tradition of visiting her sister every three months at significant expense and effort. She admitted to "only a little" fear of aging, but I suspected that this was a much more significant issue for her than she realized.

I diagnosed TMS. She seemed to understand the diagnosis and

initially agreed to attend my upcoming leçture. When I called to remind her to come, however, she stated she was getting a second opinion from another orthopedic surgeon. I accepted this and wished her well with her second opinion.

I subsequently found out through a mutual friend that her new orthopedic surgeon performed shoulder and "tennis elbow" injections. While the injections seemed to produce nice improvement initially, the symptoms recurred within a couple of weeks. I subsequently discovered that she was in the process of finding yet another orthopedic surgeon for yet another opinion. This woman seemed bound and determined to find someone who could "fix" her pain problem!

In retrospect, it may have been a mistake to give her the full TMS diagnosis when her symptoms had been present for only two weeks. It's only natural for patients to hope for a quick fix when symptoms are present for such a short time. In an acute presentation, I currently present the short version of what TMS is and how it relates to a patient's own presentation of pain. I state that stress, tension, anxiety, and anger can cause pain in muscles, tendons, and nerves. I provide reassurance that the condition is benign, and that they may improve without much treatment. I give these patients a TMS handout and tell them to return if they don't see improvement in a reasonable period of time.

THE LECTURE

Delivering the TMS lecture is particularly helpful to me in gaining insights into the questions people have about TMS. I find the lecture to be a good learning experience not only for patients, but also for me. I have learned to anticipate a lot of the questions and concerns that people have. I try to give the lecture something of an individual flavor by emphasizing points that will be relevant to individual patients.

The group setting is helpful to a lot of people because it lets them know they aren't the "Lone Ranger" when it comes to TMS. Meeting other people attending the lecture who have the same condition is a source of reassurance for many people. I welcome attendance and participation of spouses and family members at the lecture. Knowledge and understanding of TMS on the part of everyone close to the patient is most helpful in treating a condition that has defied treatment for most people.

Some people are unable to attend the lecture because of conflicts with work. I present them an alternative: they can watch the two-hour tape entitled *Healing Back Pain: The Mindbody Prescription Video* in my office during regular office hours. This high-quality presentation features Dr. Sarno presenting the lecture and answering questions from his own patients. After watching the video, my patients spend fifteen minutes with me in discussion of the ideas on the tape, and I answer specific questions they may have. The only objection I have heard from a few people is that the video seems somewhat like an infomercial and lacks the spontaneity and energy of a live presentation. For this reason, I strongly encourage most people to attend the lecture at the scheduled time if they can fit it into their schedule.

The atmosphere in the room at the end of the lecture is usually quite upbeat. People report a sense of hope for the first time in years. The sense of exhilaration that many people have is palpable. They realize that they are physically normal, and that they can expect healing of their chronic pain and suffering. They are given a study program and are advised to do "homework" every day. Daily reflection is important.

About a month after the lecture, I have my nurse call patients to see how they are doing. She asks if they have any questions about TMS and about their own manifestation of TMS. She asks if they are doing their homework on a daily basis, and about their recent pain scores, and whether they are still taking medication for pain. Patients are encouraged to talk with me on the phone if they have significant questions or concerns that they feel are better addressed by me.

They are instructed to return to the office within two months after the lecture if they are still having significant pain. At that follow-up visit, I try to get an idea of the specific psychological issues that appear most significant. If patients are working diligently on their homework and making good progress, we may hold off on a referral for psychotherapy, allowing a bit more time for them to see a reduction in their pain. Even though some people still report pain two months after the lecture, they usually strike me as significantly more focused on the real issues in their life. The homework process forces people to attend to the psychological pressures and stresses that we all face. All my patients seem more well adjusted and happier as a result of the education treatment.

Patients who go through the education process are much more willing to accept a referral for psychotherapy when it is indicated. They have quite a bit of insight into their own emotional issues by doing the homework. The daily study program is excellent preparation for psychotherapy. I have great respect for those patients who enter psychotherapy. It takes courage to look at issues that many people would prefer to avoid. I see several patients regularly for pain management while they are undergoing psychotherapy. They are uniformly grateful to me for pointing them in the right direction, that is, to think psychological, not physical. The major issue in their life has ceased to be their chronic pain problem, but rather some stressor or emotional issue they are dealing with in psychotherapy.

CASE STUDY: REFERRAL FOR PSYCHOTHERAPY

A forty-six-year-old nurse presented with low back pain and right thigh and knee pain for two weeks prior to her visit. She had a history of similar low back and leg pain problems going back ten years. Examination showed significant tenderness along the entire iliotibial band of her right leg, as well as the paraspinal muscles of the lumbar spine bilaterally. There was diffuse tenderness in the tendons around the periphery of the kneecap. X-rays of the knee and lumbar spine showed minimal degenerative changes.

Her social history included many difficulties. Her father was an alcoholic, and she endured significant emotional abuse from him during most of her childhood. During her first marriage she had been emotionally and physically abused for most of the twelve years of the relationship. She is a perfectionist who tends to be overly sensitive to criticism.

I diagnosed TMS, a condition she readily accepted. Her experience as a nurse provided significant insight into the relationship between stress and various medical conditions. She read the Sarno book, attended the lecture, and did her homework diligently. She experienced complete resolution in her symptoms within just two weeks.

She remained symptom free for four months before experiencing a recurrence of TMS, this time in her neck and upper back. I examined her and confirmed that her TMS had changed location. I was able to reassure her that her neck and upper back were normal. The question was: what new stressor in her life had precipitated the need for recurrent TMS symptoms?

It turned out she had been very upset by a recent revelation involving her fifteen-year-old daughter. The girl had been sexually abused some eight years previously by her father, and it was the discovery of this abuse that had led my patient to obtain a divorce. The new revelation was that her ex-husband's parents had long been aware of the sexual abuse of their granddaughter by their son. They had kept this information covered up in order to protect him!

She was enraged by this new revelation but felt powerless to express her outrage to her former in-laws. She expressed significant relief in being able to talk with me about this very upsetting information. She understood the psychological need to keep the unpleasant emotions contained in the unconscious mind. The appearance of TMS in a different location was a manifestation of location substitution, the symptom imperative.

I reviewed her social history with her, and we both agreed that insight psychotherapy was indicated because of the depth of unpleasant emotions she had generated over the years. I explained that it was likely she would have other recurrences of TMS if she did not enter psychotherapy. She gladly accepted the referral, confident in the TMS diagnosis because of her prior positive response to the education treatment. She was also able to arrange psychotherapy for her daughter, who has been having very significant headache pain recently.

CASE STUDIES

Carpal Tunnel Syndrome

A fifty-one-year-old man was referred by another physician for bilateral carpal tunnel syndrome. He brought with him a copy of a nerve conduction velocity test that had been ordered by the referring doctor. The test was positive for carpal tunnel syndrome. He attributed the symptoms to strenuous manual work he had been doing for many years as an electrician. He also had a thirty-year history of chronic neck and back pain, but they were not bothering him at the time. I knew I was probably dealing with TMS, but I decided to go along with the diagnosis of carpal tunnel syndrome. After all, both he and his referring physician expected me to do surgery because I am a surgeon! I performed carpal tunnel releases on both wrists on the same day. The operation went well, and he seemed much improved in the early postoperative phase.

He came back three months after surgery, however, and reported increasing tingling and numbness in both hands. He reported an increase in the pain to a level only slightly less than it had been before surgery. He wondered why the operation had provided only temporary and partial benefit. Interestingly, his neck and shoulders had recently become quite painful. The areas of tenderness in the neck and shoulders were consistent with TMS. He had areas of tenderness in the lumbar paraspinal and gluteal muscles, which were also consistent with TMS, but he had only mild complaints of pain in those areas.

At this point, I obtained a complete social history. He had a lot of perfectionistic tendencies but preferred not to label himself as a perfectionist. He tried his best to please people whenever possible. He was very conscientious in the performance of his job as an electrician. He was recently divorced, with a lot of "issues" with his ex-wife.

Based on the tenderness findings and his personality profile, I diagnosed TMS. I explained that the temporary improvement following surgery was related to the placebo response. His pain, tingling, and numbness recurred because carpal tunnel syndrome had not been the most accurate diagnosis. His neck, shoulder, and back pains were also manifestations of TMS. It took some courage on my part to admit that my original diagnosis was not correct. Fortunately, he accepted the TMS diagnosis and agreed to the education treatment.

He was much improved during the first month after attending the lecture, and now, four months after, he is totally asymptomatic. He no longer experiences pain, tingling, and numbness in his hands and wrists. He reports no lumbar, cervical, or shoulder pain whatsoever. His case has proven to me that carpal tunnel syndrome is simply a presentation of TMS in the wrist.

And what of the positive nerve conduction velocity test? I have concluded that TMS can reduce the speed of nerve conduction. TMS causes mild oxygen deprivation, which can show up when a nerve conduction velocity test is performed. Other doctors usually attribute positive test results to the effects of repetitive manual gripping and squeezing activity at work. They usually consider this to be work related. However, this interpretation is incorrect.

My patient had been having lesser symptoms of TMS in the wrists for several years before surgery. The oxygen deprivation of TMS eventually became severe enough to cause an abnormal finding on the test. This case points out the fallacy of concluding that pain, tingling, and numbness result from carpal tunnel syndrome. Even though I performed the indicated surgical treatment for carpal tunnel syndrome, the results were poor. It was only after the education treatment for the correct diagnosis, TMS, that he experienced complete resolution of all his symptoms.

Fibromyalgia

A forty-five-year-old woman presented with cervical and lumbar spine pain. She attributed the pain to an automobile accident three years previously. MRIs of the neck and back showed only the normal degenerative changes one would expect at her age. Even though no surgery had been recommended, her previous orthopedic surgeon said that her chronic pain was caused by these abnormalities! She was seeing me for a second opinion on her back and neck pain.

She mentioned a prior consultation with a hand surgeon for apparent carpal tunnel syndrome in both wrists. She also reported pain in the front of both shoulders and on the outer aspect of both elbows. The hand surgeon had mentioned carpal tunnel release as a possible treatment, but he had been noncommittal about the shoulder and elbow pain. None of her physicians had mentioned fibromyalgia as a diagnostic possibility.

Her social history revealed that during her childhood she endured emotional, physical, and sexual abuse at the hands of her stepfather for several years. Her mother basically ignored her during her childhood. She readily admitted her perfectionistic personality traits and stated she was compulsive about doing things as well as she possibly could. She was currently trying to reconcile with her much older husband, who had recently been having a lengthy affair with another woman. She had a very difficult relationship with his three adult children, who were still heavily involved in their father's life.

Her medical history included skipped heartbeats and palpitations, irregular menstrual periods from endometriosis, headaches, and dyspepsia (probably GERD or gastritis, which had yet to be worked up). She had been in and out of psychotherapy for twenty years and currently was seeing a therapist in another state.

I diagnosed TMS. I stated that other doctors might possibly use the term *fibromyalgia.* Such a diagnosis is not helpful because other doctors say they don't know the cause of fibromyalgia, and they don't have anything effective to treat it. The treatments they recommend are part of a "shotgun" approach. Other doctors dismiss a psychosomatic explanation of fibromyalgia.

She readily accepted my diagnosis and treatment. I sent a letter of referral to her psychotherapist, strongly recommending that she have insight psychotherapy, as opposed to the cognitive approach she had been using. I enclosed information about TMS, and followed up with a phone call a week later. I explained to the therapist that TMS theory calls for more than a superficial, cognitive approach. It was clear in this case that some very strong emotions from childhood required deeper exploration.

Six months later, there has been very significant progress. Pain scores are much lower, the carpal tunnel symptoms are gone, and her activity level is much improved. She continues in psychotherapy, dealing with the issues from her childhood, as well as her everyday stressors. I fully expect that she will eventually have full resolution of her symptoms, although psychotherapy may be necessary for some time.

This case illustrates several important points. Most orthopedic surgeons focus on MRI findings as the cause of back and neck pain. In this case, these abnormalities were normal abnormalities. It is anatomically impossible that the reported disc degeneration in the lumbar and cervical spine could produce such diffuse tenderness findings.

Fibromyalgia is a relatively recent diagnosis, finally recognized as a distinct entity by the American Rheumatological Association in 1990. Many physicians, myself included, had no formal exposure to the diagnosis and treatment of fibromyalgia during medical school, internship, and residency. Yet fibromyalgia is now the second most common condition (after osteoarthritis) in the field of rheumatology. It is truly proliferating at epidemic proportions.

It's not surprising that many physicians do not recognize the condition when it presents to their offices. They tend to focus entirely on the chief complaint, ignoring other aspects of the overall pain presentation. Many superspecialists look at only one area of the body, such as the back, neck, or upper extremity. A pain diagram is an excellent tool in the recognition of many chronic pain conditions, because it gives an overall impression of all the areas of pain.

In this case, my patient was convinced that the automobile accident of three years ago had caused a permanent injury to her back and neck. She stated that the initial presentation of back and neck pain came to involve her shoulders, elbows, and wrists about a year after the injury. Fibromyalgia often evolves over a period of several months or even years. It can present a confusing clinical picture to different physicians who see patients at different stages in the process.

Another point concerns psychotherapy: She asked why her twenty years of intermittent psychotherapy had not resulted in resolution of her pain. When I found out that her therapist had taken a cognitive approach to all the stressful events in her day-to-day life, I stated that a deeper, insight-oriented form of psychotherapy was necessary. Her childhood had been marred by emotional, physical, and sexual abuse, resulting in a large amount of unconscious rage. This directly contributed to her perfectionism and goodism. Even though some of her earlier therapy had explored these issues, they had not been resolved. Her psychotherapy needed to be redirected to a deeper exploration of these unpleasant emotions in the unconscious mind.

It is quite clear to me that fibromyalgia is a psychosomatic condition. It appears to be a severe manifestation of TMS. Other authorities reject a psychosomatic explanation, preferring to say that the cause is unknown. The sad fact is that most physicians have not been trained to obtain an in-depth social history in patients with chronic pain. They do not accept the idea that unpleasant emotions in the unconscious mind actually cause physical symptoms.

Only when these basic facts are understood and accepted will we see an end to the current epidemic of TMS musculoskeletal pain in its many varieties, including the mysterious fibromyalgia.

NINE

STRUCTURAL PAIN OR PSYCHOSOMATIC PAIN?

Douglas Hoffman, M.D.

Raised in the Washington, D.C., area, **Douglas Hoffman, M.D.**, received a bachelor of science degree from Duke University and his medical degree from the University of Vermont College of Medicine. Dr. Hoffman completed a family practice residency at St. Margaret's Memorial Hospital in Pittsburgh, Pennsylvania, and a primary care sports medicine fellowship at Hennepin County Medical Center in Minneapolis, Minnesota. Currently, he practices nonoperative orthopedics/sports medicine in Duluth, Minnesota, where he is the team physician for the University of Wisconsin-Superior and the Duluth Huskies baseball team. Dr. Hoffman enjoys sports, including Nordic skiing, cycling, and kayaking.

As you will see, Dr. Hoffman is most sensitive to the psychological, social, and societal influences in psychosomatic disorders, and suffers the frustration common to everyone who does this work of finding so few patients who can profit from his expertise.

STRUCTURAL PAIN OR PSYCHOSOMATIC PAIN?

Not long ago, a family physician traveled several hours to see me for his chronic low back pain. He had read Dr. Sarno's book, *The Mindbody Prescription,* after the usual "standard" medical treatments had failed to relieve his symptoms. Our one-hour visit together consisted of hearing the history of his pain, learning about his life, and helping him to understand and reinforce the psychosomatic nature of his back pain. As we both expected, his symptoms completely resolved over the next few weeks. I recall his asking me as we were walking out of the office after his appointment, "How can I go back and practice the way I've been practicing knowing what I know now?"

It's an important question, and he's not alone in asking it. I live a dual existence in my nonoperative orthopedic practice: I treat patients' musculoskeletal pain problems by the current standard of care, with the rare exceptions of those who are willing to see their problems as psychosomatically induced.

Let me describe a few patients from a typical morning in my office, starting with a fifty-two-year-old woman who came in with severe right hip pain. She had extreme pain with hand pressure (palpation), medically referred to as *tenderness*, over the right side of the right hip, tenderness of the right upper buttock, and along the course of a long tendon on the outer aspect of the right thigh. Similar corresponding landmarks were tender in the other leg, but less so. She had a history of depression, migraine headache, obesity, upper gastrointestinal reflux, high blood pressure, foot pain (plantar fasciitis), and carpal tunnel syndrome, and she had had a hysterectomy for a uterine fibroid. X-rays of the hip were normal. The diagnosis was greater trochanteric bursitis. She received standard care.

A thirty-five-year-old woman, a certified nursing assistant at a local nursing home, was moving a patient two weeks ago when she developed pain, numbness, and tingling on top of the right shoulder, the right shoulder joint, and the right arm. Her pain was so severe that the combination of a muscle relaxant, an ibuprofen-type drug, and a narcotic resulted in only minimal improvement in the pain. On examination there was diffuse pain to mild palpation of the involved areas. Her past medical history included depression, whiplash injury, migraine headache, and painful menses. She was a single mother of two and smoked a pack of cigarettes a day. She was treated with conventional methods for pain.

A forty-year-old man returned for a six-week follow-up of an ankle fracture, treated with a cast. He complained of severe, persistent pain and had been taking narcotic painkillers since the cast was put on. Out of the cast, there was no evidence of swelling but he had severe pain on palpation all over the ankle and would not allow any movement of the ankle. X-rays showed good healing. He had a history of alcoholism, bipolar disorder, and chronic low back pain, for which he was on Social Security disability. He, too, received standard pain treatment.

A sixty-five-year-old recently retired man came in with right knee pain. I had previously treated him for arthritic changes in the left knee, which eventually required a total knee replacement due to intractable pain. There was a history of two low-back fusions, rotator cuff repairs at both shoulders, bilateral carpal tunnel releases, and recent surgery for an enlarged prostate. He awakened one morning and the pain was there. An x-ray showed moderate osteoarthritis (normal aging changes). Treatment was standard.

Although it is likely that all four of these patients were suffering from psychosomatic pain, it was my clinical judgment that they would not be accepting of that diagnosis.

The last patient of the morning was a forty-two-year-old woman who traveled several hours to see me for her chronic neck pain of ten years. A year ago her pain worsened, prompting a medical evaluation that included an MRI and orthopedic and neurosurgical consultations. Surgery was not recommended. Chiropractic treatments, various analgesics, and working with a personal trainer all resulted in only temporary improvement. She had read Dr. Sarno's book, *The Mindbody Prescription*, a year ago, again with temporary improvement. She found my name on a related website and called for an appointment. During our one-hour visit, we discussed the psychosomatic nature of her pain, including its cause, possible precipitating factors, and techniques to effect a cure. A follow-up letter and phone call revealed that she is now predominantly pain free. The following is the last paragraph of her letter: "I am finishing this letter after having been outside gardening for two hours. At times I doubted I would ever garden again."

How did I come to practice medicine with a dual existence? It stems from my own personal history. I can remember vividly the first time I experienced low back pain. As a medical student, life's pressures were building, as was my drive for perfection. This drive also included bicycle racing. It was on a typical training ride that I first experienced tightness in my lower back. But this wasn't just an ordinary training ride. I was trying to set a personal time record. The pain was so severe I was unable to continue bike racing. I underwent an evaluation by a physician who subsequently referred me to a physical therapist. Despite my dedication to daily exercises there was no improvement.

In the subsequent years my pain gradually worsened, and each physician I sought treatment from had a different diagnosis. Each physical therapist also had a different diagnosis and a different solution to the problem. But the pain in the lower back prevailed, and I was relegated to leisurely bicycle rides. It was on one of these rides before graduation from medical school that I came upon one of the faculty orthopedic surgeons also out for a ride. After hearing my story, he simply said, "You're probably just one of those Low Backers."

Unaware of how those powerful words programmed me to experience ongoing back pain, I continued to fail various treatment attempts with both conventional and alternative approaches. I was desperate! One day at the bookstore I saw Dr. Sarno's second book, *Healing Back Pain: The Mind-Body Connection.* How true were the words that I read. What perfect sense he made. Applying the theories and treatment techniques that I learned from the book, within one week my back pain was better than it had been in ten years and in six weeks I was "cured." I was stunned! Simply by understanding my pain as psychosomatic, including its psychological reasons, and "undoing" all the ways in which I was programmed to have back pain, I was pain free!

Wait a minute—were all those physicians and physical therapists I saw, all my medical school mentors and medical textbooks wrong? I asked myself how was this new knowledge going to affect my practice.

My training is in family practice, primary care sports medicine, and nonoperative orthopedics. The majority of my practice is treating patients with musculoskeletal disorders. The year following my own cure from a chronic pain disorder, I observed and experimented with Dr. Sarno's theory in my practice. And still to my surprise, I found that what he theorized was true. Many of the common musculoskeletal disorders are not well explained by our conventional teaching and are frequently psychosomatic in origin. I wanted to learn more, so I went to New York and worked with Dr. Sarno, picking his brain and discussing his theories. After a week together, his parting words to me were those of encouragement to retrace some of his footsteps and make my own observations. The following thoughts are the culmination of nearly eight years of experience since working with Dr. Sarno, observing and treating patients with musculoskeletal disorders.

Medicine teaches us that pain, particularly musculoskeletal pain, is a consequence of an abnormality of a particular structure. This is most easily illustrated and accepted with acute trauma. When one falls on an outstretched hand and sustains a wrist fracture, the resulting structural abnormality causes pain. When the bone heals, the pain goes away. Similarly, when a track sprinter sustains an acute tear to his hamstring muscle, it hurts. There is usually swelling and bleeding as a result of the tear. But as the muscle begins to heal, the pain subsides and the athlete is back sprinting again once the healing and rehabilitation are complete.

This understanding as the origin of pain, however, is generally applied to all musculoskeletal pain syndromes. For example, after a hit-from-behind car accident people assume that muscles and other soft tissues have been damaged (whiplash syndrome). Other common examples are leg pain, referred to as sciatica and routinely attributed by doctors and patients to compression by a lumbar herniated disc; low back pain blamed on normal aging changes of discs and spinal bones; and carpal tunnel syndrome.

The term *inflammation* is commonly used to explain pain despite the fact that there is no scientific proof that it exists. A persistently painful muscle or tendon is often said to be inflamed or strained. Pain on the bottom of the foot (plantar fasciitis), elbow pain (tennis elbow), shoulder pain (impingement), and hip pain (trochanteric bursitis) are common examples of psychosomatic tendon pain attributed to some structural abnormality (inflamed, torn, worn, strained, degenerated). I refer to this as the structural model. Though there are some conditions that can be explained by this model, a careful study of common nontraumatic pain complaints reveals that most are not structurally induced. However, structural abnormalities are almost universally accepted as the cause of most musculoskeletal pain, while psychological factors are thought to have little or no role in the pain, except to make it worse or better. To modify the pain is one thing, but the overwhelming majority of doctors and patients do not believe that the mind can actually cause a physical disorder, which is the definition of psychosomatic.

In the course of my work I have been greatly impressed with the common misperceptions about the definition of a psychosomatic disorder. Grossly mistaken ideas, like "the pain is in your head," or "it's imaginary," or "the person is a hypochondriac (or mentally unsound)," are universal. Even physicians are guilty of harboring such opinions. If one considers the disorder that caused my back pain, TMS, and all its equivalents (stomach, colon, allergic, dermatologic, etc.), the fact that we experience them when we're nervous, when we're sad or glad or sexually aroused, it is clear that psychosomatic reactions are universal, normal, and part of the human condition.

It is not unusual to develop a viral illness (the common cold) during a time of emotional stress, the result of weakening of the immune system. We are talking about physiologic changes that occur in various tissues or organs in the body that are induced by emotions. That is the meaning of psychosomatic.

Another thing that has impressed me is the almost universal lack of awareness in medicine where psychosomatic disorders are concerned that if you don't treat the cause of a symptom you cannot expect a cure. The nonsurgical treatment of pain is almost totally symptomatic. Pain clinics treat pain, sometimes relieving the pain thanks to a placebo reaction (cure through blind faith), but in most cases the pain returns, and so the pain epidemics continue.

Surgeons believe they are treating the cause when they perform surgery for herniated discs, spinal stenosis, malalignments, and the like; but since these abnormalities are usually not the cause of the pain, surgery will either fail or the patient may have a placebo cure. This possible surgical outcome was described in the medical literature by a Harvard professor forty-four years ago. Surgery is a powerful placebo.

Placebo cures (surgical or nonsurgical) are poor medicine because they do not treat the cause. If the results were permanent, we could live with that, but they are not. If the pain is relieved, one of three things will happen: either it will come back, the brain will locate the pain somewhere else, or the brain will choose another organ or system to produce symptoms. Sometimes it will substitute an emotional reaction such as anxiety or depression. I have referred to this as the *equivalency response*. This extremely important phenomenon has been described by Dr. Sarno as the symptom imperative. It means that you must eliminate a symptom by treating the cause, or the brain will simply find another symptom. According to Dr. Sarno, Freud described symptom substitution a hundred years ago but did not know what it meant.

As Dr. Sarno has pointed out, the medical profession is largely responsible for the pain epidemics sweeping the country today because it is unaware of the existence of psychosomatic disorders and ignores the possibility that much of its treatment success can be attributed to the placebo reaction.

Mrs. T's case is another example of the phenomenon. She is a sixty-year-old female intensive care nurse with a medical history that includes depression, migraine headaches, tobacco abuse, and previous lumbar back and carpal tunnel surgery. She developed severe right shoulder pain after lifting a patient at work. The pain was located throughout her whole shoulder, with radiation of both pain and numbness down her right arm. Her symptoms initially improved with a prescription for an anti-inflammatory medication, a cortisone injection, and physical therapy. When the pain became severe again, a magnetic resonance image (MRI) of the shoulder revealed a partial rotator cuff tear. She underwent operative intervention and her pain resolved after convalescence. A month later she missed a week of work due to a severe exacerbation of her migraine headaches.

Mrs. T illustrates some of the issues just discussed. She is a good example of the *symptom imperative.* Though her symptoms in the shoulder and arm were not typical of a tendon tear in the shoulder, the MRI showed a "rotator cuff tear," and despite the fact that the medical literature reports that such tears are common with advancing age and without pain, she had surgery anyway. There was a placebo "cure" but, as could have been expected, the symptom imperative kicked in a month later.

Let me share with you some of my observations in working with psychosomatic pain disorders.

THE ROLE OF TRIGGERS IN THE PSYCHOSOMATIC PROCESS

Because neither patients nor doctors recognize the reality of psychosomatic pain, they are under the impression that pain is always caused by structural-physical phenomena. The occurrence of physical incidents associated with the onset of pain reinforces for patients the idea that they have hurt themselves or that the continuation of pain after a legitimate injury (like the man with the broken ankle mentioned earlier) is still due to the injury. I have seen countless patients who fall into this category, many of them receiving Social Security disability payments. These people are not malingerers. They are unaware that the injury served as a trigger, setting off the TMS response. They truly believe that the original injury is responsible for their chronic pain.

The psychosomatic response (TMS where pain is concerned) may begin slowly and insidiously or at the time of a physical trigger. The latter may be something as mundane as swinging a golf club or lifting a laundry basket or following a legitimate injury. Dr. Sarno is inclined to limit the use of the word *trigger* to the physical incidents just described; but I have been so impressed by the role of a large number of psychosocial factors that I feel constrained to give these a special place in the etiology of psychosomatic disorders.

From an epidemiologic perspective, physical triggers are central in shaping the nature of psychosomatic disorders. The unconscious mind strives to present symptoms that are legitimized in society. It wants to be taken seriously! Therefore, a pain disorder arising from a physical trigger serves to legitimize the psychosomatic one. For example, carpal tunnel syndrome may occur from a structural cause such as pregnancy, hypothyroidism, or uncontrolled diabetes. But these are not common causes, and furthermore it does not explain the epidemic proportion that carpal tunnel has occurred recently in our society. Thus, the carpal tunnel syndrome from a structural cause (an uncommon occurrence) becomes the template for the disorder to occur with relative frequency as a psychosomatic disorder.

Before the eradication of polio in the United States, "hysterical paralysis" was a common diagnosis for what was, in fact, a psychogenic (conversion-hysterical) disorder. The paralysis resulting from polio (a structural cause) was the template for the psychogenic paralysis. But when polio was eradicated through the advent of a vaccine, hysterical paralysis became exceedingly rare. During large-scale chemical or toxic exposures such as from war or industrial accidents, many patients who did not have a direct exposure will show up with identical symptoms to those who did. Thus, as the nature of structural illness changes over time and between cultures, so does psychogenic illness.

On a more individual level, physical triggers are one of the most common catalysts of the psychosomatic process. Automobile accidents, falls, physical abuse, and repetitive motion in the workplace are examples of physical triggers that often result in chronic pain syndromes.

Mr. S is a forty-two-year-old construction worker who slipped on the ice and fell backward, landing on his back, while at work. He experienced immediate moderate low back pain, but over the next several days his pain worsened to the point where he had trouble walking and sleeping despite pain medication. He continued to have constant and severe low back pain, and multiple medical evaluations showed no acute traumatic injury. After nearly a year of pain and inability to return to work, he was under the care of a pain specialist with chronic narcotics. Two years after that injury he was placed on Social Security disability.

The above case exemplifies how a physical stimulus (contusion to the back from a fall) triggered a psychosomatic process (chronic low back pain). His medical evaluation ruled out the presence of an underlying bone fracture. One would expect to have a bruised and sore back for one to two weeks from a fall, not a chronic pain syndrome.

The propensity to trigger the psychosomatic process is not only related to the physical stimulus but also the environment in which the physical trigger occurs. For example, a motor vehicle accident resulting in whiplash syndrome involves a physical trigger as well as cultural, legal, and insurance influences.

Ms. R, twenty-two years old, was a belted driver involved in a rear-end motor vehicle accident. While her car was at a stoplight, another car hit her from behind going about twenty miles per hour. For several days she had mild neck stiffness, but a week after the accident she awoke with severe neck pain and stiffness with associated headaches. Several days after the accident she also received a phone call from the insurance company of the driver who hit her from behind, offering to "settle the case." She refused and contacted an attorney to represent her. Due to worsening symptoms she was evaluated in the emergency room, which included a normal MRI of the head and neck. One year after the accident, she still experienced neck pain and headaches despite ongoing treatments. Litigation remains pending.

The above case is commonplace in our society and exemplifies how a physical trigger, such as a rear-end accident, results in a chronic pain syndrome through the power of multiple psychogenic influences. Culturally, the whiplash syndrome is widely accepted and legitimized in our society, as is prolonged disability from an accident. Furthermore, our society places a dollar value on the injury by way of the legal and insurance industries, which in turn shape our values regarding disability and the victim mentality (not necessarily consciously). Finally, the media further strengthens these attitudes through the advertising of accident attorneys, coverage of litigation, and advertising of various medical treatments. In effect, the media heightens the awareness of the injury, which, ironically, increases the likelihood that a chronic whiplash syndrome will occur with rear-end accidents. This awareness of a disorder, as I mentioned above, is a strong trigger to developing a psychosomatic disorder.

Similarly, when societal influences are altered or eliminated, then the psychosomatic process will subsequently change. One such study that illustrates this was published in the *New England Journal of Medicine* in 2000. Noting a rise in the number and cost of whiplash claims in Saskatchewan, Canada, and the lack of objective data regarding its cause, the province of Saskatchewan changed the government insurance from a tort-compensation, which includes payments for pain and suffering, to a no-fault system, which does not include such payments. This change went into effect on January 1, 1995. Not surprisingly, there was a marked reduction in both the incidence and recovery time of whiplash injury when comparing those that occurred six months before to six months after the insurance change.

Physical abuse is another dramatic example of how a physical stimulus often results in a chronic pain syndrome of psychogenic origin. It is not unusual for a contusion or fracture as a result of physical abuse to result in chronic pain despite apparent healing from objective criteria. This is an example of a physical trigger interacting with an emotional one to produce a psychosomatic disorder. The challenge is to determine when a physical trigger results in a structural disorder or a psychosomatic disorder. I will discuss this later.

Societal influences are numerous and quite powerful in shaping the psychosomatic process, not only for a given individual but also for a society as a whole. Culturally, we are a society that promotes and rewards those individuals who blunt their emotions. Being "cool, calm, and collected" is a positive value in our culture. Parents often tell a child who is feeling pain, whether physical or emotional, that "everything is okay" or "you'll be all right" rather than hug them and allow them to express their distress and thereby validate what they are feeling. Our society is emotionally well defended. However, the difference between how we think we should feel about a situation or event in our lives (our conscious emotion) and what we really feel (often the unconscious emotion) becomes psychogenic.

When a patient says to me "I'm not going to let it (stress) get to me," red flags go up. It is not uncommon for a patient who chooses to treat their physical disorder through Dr. Sarno's methods to begin to experience life in a more genuine way—in other words, to become more aware of emotions as they really occur. However, many individuals would find living life in a more genuine manner more difficult than the physical pain they are experiencing because they would have to acknowledge painful parts of themselves or painful emotions buried deep inside their unconscious mind. Theoretically, in a society that encourages and supports expressing emotions, there would be less of a need for the psychosomatic process to supersede.

The media is another important cultural influence shaping the psychosomatic process. The general population is exposed to ever increasing amounts of health care information and advertising, unparalleled in history. As a consequence, people are as likely, or in some instances even more likely, to believe the information they gain about health issues through the media than from health care professionals. There has been an unintended shift of authority from the physician to the media. For example, the intense advertising of a new drug by a pharmaceutical company will often increase the "power of placebo." With every new nonsteroidal anti-inflammatory drug (NSAID) that has hit the market, I have observed a much greater efficacy for approximately one to two years before its therapeutic value decreases to the level of all the other ones available. Another example is the intense advertising of alternative medicine, particularly that which aims at low back pain. From magnets to mattresses, supplements, traction devices, shoe inserts, and even spring-loaded cushioned shoes, a lucrative industry has developed in search of the quick fix. Unfortunately, patients will experience only a temporary improvement (or an equivalency reaction) with these "placebo" devices. Finally, the media's affinity for controversy can be misleading to the public with statements such as "The [White House] promised relief today to millions of workers with aching backs, crippled fingers, sore wrists, and other physical problems caused or aggravated by their jobs." This type of statement naturally leads the public to believe that repetitive activity causes incapacitating musculoskeletal disorders.

The legal system in today's society is also a strong trigger of the psychosomatic process. Our society places a dollar value on suffering, which becomes psychogenic (unconsciously, of course) in and of itself. Furthermore, there is intense advertising among attorneys to attract business, whether from an automobile accident, work injury, or injury on someone else's property. This fuels the notion that someone else is to blame for our injuries, so we are the victims. The courts have, in general, supported and legitimized this way of thinking. Consequently, when an injury does occur and the legal process is engaged, both become strong triggers for the development of a psychosomatic disorder. Again, the whiplash syndrome is a common example where the legal system is engaged (with the subliminal message that you are a victim and your pain is worth money); the medical system is involved, including a variety of therapeutic disciplines, which further validates the "injury"; and the insurance industry is involved, holding the money and arguing over who should pay.

Important legal decisions also shape the way society views injury, particularly in the workplace. This certainly was the case in Australia, where a precipitous rise in the rate of work-related repetitive strain injuries began to occur in 1983. The single greatest influence that reversed this epidemic was a judicial decision in 1987,

Cooper v. the Commonwealth, in which the Australian Supreme Court ruled that the employer was not guilty of negligence and the plaintiff had not suffered an injury. All costs were levied against the plaintiff, and subsequently the repetitive stress injury disappeared. We are in dire need of a similar judgment in the United States.

Similarly, the insurance industry influences the psychogenic process, because it increases the awareness and further legitimizes pain syndromes resulting from injury. Furthermore, it reinforces the victim mentality as well as the monetary value of an injury. It often evokes anger and resentment with one's injury (which in turn evokes unconscious emotions) with the common practice of denying claims or requiring engagement of the legal process to avoid paying money.

A subsection of the insurance industry, the workers' compensation system, is one of the strongest psychogenic influences of musculoskeletal problems in our country. At one time in our country's industrial history, protection of workers from dangerous work conditions and exploitation was desperately needed and was central in empowering the worker. Even today, laws that protect workers are very important and needed in many circumstances. However, conditions for workers have generally changed, and the pendulum has swung in the other direction. Any type of musculoskeletal pain syndrome that develops in association with work is considered a workers' compensation injury. This, in turn, influences the way society thinks about any type of work-related pain that triggers the psychosomatic process. The inability to work because of a medical condition, particularly a musculoskeletal disorder, has become more socially acceptable over the last several decades. In parallel, the incidence of work-related disorders such as low back pain and carpal tunnel syndrome has risen in epidemic proportions, despite the fact that workers are subjecting themselves to less repetitive motion and strain in the workplace. This paradox is explained, in part, by understanding that the workers' compensation system itself is a psychogenic trigger for developing a pain disorder.

Finally, the health care industry plays a big role in validating and perpetuating psychosomatic disorders. As I mentioned earlier, medical school training and, therefore, medical practice does not understand or acknowledge psychosomatic disorders. Thus, physicians can only explain pain based on the structural model. This has several important implications. First, treatment strategies can only be derived from this model. Second, validating one's psychosomatic pain disorder as a structural problem reinforces the reason the pain is there in the first place, to distract one's unconscious emotions. Thus, current medical treatments often reinforce the psychosomatic process because they legitimize and validate the pain as structural. If physicians did not validate psychosomatic problems as structural (e.g., fibromyalgia), then the problem would diminish, or even disappear.

Why are physicians so reluctant to accept such a concept? The answer is complex and includes our medical training biases as well as *our own unconscious emotions*. To be able to understand and effectively apply Dr. Sarno's principles, one must acknowledge their own shadows and painful emotions. Additionally, a physician's position of authority is compromised with this perspective since patients become empowered by discovering *that they have the ability to heal themselves*.

Emotional triggers occur when circumstances, whether dramatic or more subtle, evoke unconscious painful emotions and fuel the need to defend them with a distraction such as physical pain. In other words, real stress is that which evokes unconscious emotions. Life cycle events such as marriage, having children, moving, aging, and growing older are often strong emotional triggers. It is not uncommon for a woman to develop neck or shoulder pain from holding her infant or toddler. This is an example of both a physical trigger (holding the infant) and an emotional one (stress of motherhood, possibly evoking one's own childhood memories).

A forty-four-year-old male physician was getting dressed for his daily morning run. He anticipated a shorter run that particular morning since this was his first day in a new faculty position and he wanted to arrive early. Upon bending down to tie his shoe he developed sharp, intense pain in his lower back. He was able to go to work, but his pain did not subside for several days. Six months later he still had aching in his low back and had not returned to running for fear it would worsen again.

This physician had bent down to tie his shoe thousands of times without difficulty. What was different this time? He was experiencing emotional stress in anticipation of his first day at a new job. Not just the conscious stress of starting a new job but the unconscious emotions that were probably evoked, such as feelings of fear and insecurity.

A thirty-two-year-old female attorney developed low back pain several years ago associated with gardening. A physician evaluation revealed mild degenerative disk disease of the lumbar spine. There was no improvement despite medication, chiropractic treatments, and physical therapy. She had also explored complementary (alternative) treatments without success.

Her personal history was significant. She had a difficult relationship with her mother, who she described as "controlling." Throughout high school and college she was a straight-A student and graduated at the top of her class at a nationally prestigious law school. She had also been undergoing evaluation and treatment for infertility. Several years before, she had quit her successful law practice with a private firm and started working in a less demanding position for a nonprofit organization. Due to her continuing troubled relationship with her mother, she had initiated psychotherapy several times in the past three years.

This case illustrates several points. First, a controlling parent can be enraging to a child, which is one of a number of painful unconscious emotions that this patient has been protecting. Furthermore, her academic achievements and success as an attorney stem from the parts of her personality that strive for perfection, emotional control, and independence, which have served her well and have often been rewarded. However, in the unconscious mind her perfectionism may also be enraging. Her natural tendency to avoid emotions can signal an unconscious fear that if she allows her repressed emotions to go unprotected, they may come pouring out uncontrollably. Finally, for much of her young adult life her distraction from these painful unconscious emotions has been in the form of achievement in school and subsequently a busy law practice. There was no need for physical pain since she already had plenty of distractions.

However, when she quit her busy practice she lost an important defense mechanism. She also began to experience more stress in the life cycle event of marriage. I would also imagine that dealing with infertility was quite psychogenic for several reasons. It is in her nature not to express grief (over her inability to conceive a child) but to repress it, adding to the repository of those painful emotions that lie in the unconscious. Also, infertility is often out of the control of the person going through it. This can also be enraging, especially for an individual who has learned that if you work hard enough, your rewards will come. So, with the building of life stresses and the loss of distractions, the back pain saved the day and kept those emotions repressed.

Dr. Sarno has emphasized the important role that pain plays in distracting one's mind away from the emotional to the physical. I find that there are other pain equivalents that serve as distractions, one of the most common being the workaholic ethic. I can remember a coworker telling me, "Why would I ever want to take a vacation; I wouldn't know what to do with myself!" In other words, unconsciously he is saying, "Why would I ever want to sit with my emotions?" It would be too painful for him!

In my practice one of the most common life events that may trigger the psychosomatic process is the process of aging. Aging often encompasses such issues as failing health, loss of independence, and the realities of mortality—both in oneself and in loved ones. The evaluation of these patients is challenging since the incidence of other structural diagnoses such as cancer, organ failure, neurologic, and rheumatologic disorders is higher in the elderly population.

PAINFUL UNCONSCIOUS EMOTIONS FUEL THE PSYCHOSOMATIC PROCESS

There is no question that unconscious emotions are the foundation of the psychosomatic process. It has been my experience that any painful unconscious emotion can be psychogenic since it can be threatening to us and thus we naturally want to defend against it. Anger or rage is a common emotion that warrants our defenses but usually is a response to repressed emotions such as shame, guilt, feeling unwanted or inadequate, sorrow, fear, and insecurity.

A thirty-six-year-old woman with a past medical history of migraine headaches and an anxiety disorder developed neck and bilateral shoulder pain four months after the delivery of her first child. She attributed the pain to holding her son. It improved after a course of physical therapy, which was about the same time she returned to work as an accountant. Eighteen months later she gave birth to a second son and decided to quit her job and stay home with her children. She began experiencing intermittent neck pain again and also had a marked worsening of her migraine headaches. Six months later, her mother, who had been diagnosed with lung cancer, died unexpectedly. Her mother had also been an alcoholic, and the patient had experienced frequent outbursts of anger from her mother as a child. She had also become increasingly concerned about her husband's drinking, although she had not discussed it with him or anyone else. Three weeks after her mother's funeral she developed increasing joint pain, generalized aching, and fatigue. She hurt everywhere! After several physician visits she was diagnosed by a rheumatologist with fibromyalgia. Due to the disabling nature of her pain, she required child care during the day.

This patient harbors many difficult emotions related to her childhood, including emotional abuse from an alcoholic mother. These repressed emotions along with the pressures of becoming a new mother triggered her neck and shoulder pain that improved with the distraction of going back to work. The added pressures of a second child combined with the decision to stay at home with them stirred up her repressed emotions further as evidenced by the return of neck pain and the worsening of her migraines. Finally, her mother's death became too much! Not only did she lose a mother but her death triggered many of the repressed emotions stemming from her childhood emotional abuse and her mother's alcoholism. The icing on the cake was her husband's drinking. Unconsciously, her evoked emotions felt like they would be too much to handle (rightly so) and thus required drastic symptoms to adequately defend against their threat of expression.

Patients who were physically, emotionally, or sexually abused often experience a lifetime of physical pain rather than having to grieve and experience the unbearable emotional pain that became repressed in childhood from their awful experience. This is an example of the symptom imperative: the need to have physical symptoms, often and ongoing, in order to keep one's emotions repressed and well defended. To emphasize again, I have found that the severity and chronicity of one's symptoms is proportional to the underlying pain that harbors in the unconscious and the triggers that evoke them.

It is common for patients or physicians who understand the true cause and thus the appropriate treatment of psychosomatic disorders to want to tell their friends or acquaintances about their disorder and how to successfully treat it. They are often met with resistance. I can recall giving a talk about the mindbody connection of fibromyalgia to a group of 400 people with fibromyalgia, a particularly severe manifestation of TMS. These were people whose symptoms were disabling and consuming, but not one person was willing to see their problem as psychosomatic! Ironically, this makes perfect sense. If one acknowledges that his disorder is psychogenic, then he is acknowledging those painful unconscious emotions. He is shedding his defenses and, thus, his distractions. Consequently, he would begin to feel emotional pain! For many, this would be too much. It would be easier to have their physical pain than to begin to feel those emotions that have been repressed for good reason. Of course, this thought process is not conscious. I will tell those patients that I treat with Dr. Sarno's methods that it may be an emotionally painful process. It's not easy to acknowledge that these powerful and painful emotions reside within us. That's why they have been repressed. I vividly remember a patient that I treated for chronic low back pain several years ago. She had great insight into the concept that her pain was distracting her thoughts away from those painful unconscious emotions. However, when her back pain resolved she was, as she described it, an emotional wreck. She was able to flip the switch between having back pain or struggling with her emotions. In the end, it was easier to have back pain!

We should not forget the power of these emotions. For many, it is easier to suffer with physical pain than it is to acknowledge the emotional ones. Furthermore, our society is not supportive of expressing those painful emotions we all harbor. As I noted before, we are a society that is uncomfortable with crying or grieving, and view those who become overtaken with emotion as weak. It's no wonder we often feel alone with our emotions and desire to keep them repressed. There is often no support to help grieve life's disappointments, challenges, and losses that we all experience as human beings.

HOW DOES ONE DISTINGUISH BETWEEN A STRUCTURAL DISORDER AND A PSYCHOSOMATIC DISORDER?

This is a very important question that I carefully consider with each patient I encounter because, as I stated earlier, in understanding the cause of a patient's symptoms there is the possibility of a cure. There is not a blood test, x-ray, MRI, or other form of medical technology that establishes the diagnosis of a psychosomatic disorder. In addition, a psychosomatic diagnosis is not a diagnosis made because nothing else fits; it has its own set of signs and symptoms, some of which do overlap with structural disorders. The best tool a physician can use is a thorough history and physical examination. Once this is accomplished a differential diagnosis is established, which is a list of possible disorders that fit with the presenting signs and symptoms. Technology, such as laboratory or imaging studies, may become necessary to narrow down the differential diagnosis.

As an example, a forty-five-year-old male with a past medical history of depression, gastroesophageal reflux disease, and alcoholism came to my office with right hip pain that was located over the lateral aspect of the hip and ached all the time. He developed his pain while undergoing inpatient detoxification and treatment for his alcoholism. His physical examination revealed extreme tenderness over the gluteal muscles and greater trochanter; pain was brought out with hip range of motion. Examination of the lumbar spine was normal. An x-ray and MRI of the hip were normal.

In this case the history and physical examination suggested possible hip joint disease versus a regional soft tissue pain syndrome, a predominantly psychosomatic disorder. Imaging studies were used to rule out hip joint disease. This was an important step since the incidence of hip joint disease such as avascular necrosis of the femoral head is higher in alcoholics. Review of his medical history suggested possible previous psychosomatic equivalents, including addiction, depression, and reflux. Finally, the timing of his pain (i.e., during detoxification) lent further support to the possibility that his pain was psychosomatic in origin. When this patient lost the "distraction" of alcohol use and was experiencing the stress of detoxification, he developed a pain syndrome.

There are a number of signs and symptoms that are helpful in establishing the presence of a psychosomatic disorder:

Patient's History

- A past history of other psychosomatic disorders
- Presence of a current psychosomatic disorder
- Circumstances surrounding the onset of symptoms
- History of depression and anxiety
- Timing of symptoms that suggests an equivalency response
- Symptoms that are not well explained anatomically
- Migration of symptoms
- Intermittent symptoms
- Location of symptoms

Physical Examination

- Presence of muscle trigger points
- Hypersensitivity to palpation of soft tissue
- Neurologic dysfunction (paresthesias, weakness) in a non-anatomic distribution
- Exaggerated response to pain

Not one of the signs or symptoms listed above is individually specific for a psychosomatic disorder. Rather, one must place them in context with the presenting problem as well as consideration of other structural disorders that may also explain the presenting complaints. Two examples help illustrate how a doctor may come to understand a patient's disorder.

The first case is that of a sixteen-year-old girl who developed an insidious onset of neck pain a year before while playing hockey. She was hit from behind, causing an extension-flexion-type injury with immediate neck pain. After several weeks, the pain improved but never resolved. Her pain worsened the following summer while she was playing basketball. There were no associated neurologic symptoms. Radiographs and an MRI were performed by her regular physician and were normal. She had no improvement from physical therapy. She was frequently awakened at night with the pain. Her past medical history was unremarkable. On examination she had pain over the spinous processes extending from the third to the sixth cervical levels and pain along the right neck musculature. Her range of motion was limited and painful. A bone scan revealed uptake at the fourth cervical level and a subsequent computed tomography (CT) scan showed an osteoid osteoma, a benign but painful bone tumor. Her symptoms completely resolved after surgical removal.

This case illustrates the importance of considering a differential diagnosis with her presenting complaints. While her symptoms could have been explained by a psychosomatic disorder, the age of the patient, absence of other medical problems, seemingly ordinary life without obvious family dysfunction, and consistent symptoms warranted a further workup for a structural disorder. Osteoid osteomas often become symptomatic during the teenage years, and the spine is a common location.

The second case concerned a seventeen-year-old young woman who was an unbelted driver hit from behind while stopped at a stop sign. She initially felt fine but two days later awoke with severe neck pain and headaches. An evaluation in the emergency room included x-rays of the neck and an MRI of both the head and neck, which were negative. She complained of pain and numbness down both arms and hands and had frequent headaches. Examination in the office showed marked pain with light palpation diffusely of the neck and shoulder musculature and a normal neurologic examination of the upper extremities. Her past medical history included attention deficit disorder, depression, migraine headaches, and chronic bilateral knee pain, precluding physical education class in school. Further history revealed poor school performance and drug and alcohol use. Her father left home when she was an infant and the patient never met him. She lives with her mother, who has had numerous male relationships.

In contrast to the first patient with neck pain, this patient has a number of factors that point toward a psychosomatic etiology of her neck pain. First, her pain developed two days after the accident. If tissue is acutely injured, it is usually immediately painful. Her history and examination showed widespread muscular pain and subjective neurologic symptoms that did not conform to an anatomic distribution. Her history revealed previous psychosomatic disorders, and her poor school performance, as well as drug and alcohol use, are probably an expression of her underlying painful unconscious emotions such as rage, feeling unloved, lack of nurturing, and sorrow. Her dysfunctional social situation further lends support to the likelihood that her pain is psychosomatic.

I discussed previously that a structural abnormality is a common trigger to the psychosomatic process. There are many instances in which I believe there is a spectrum between the psychosomatic and structural as the source of pain. For example, a disorder such as fibromyalgia is clearly psychosomatic in origin. In contrast, pain from a wrist fracture is due to a structural abnormality. Take a common disorder such as osteoarthritis of the knee, for example. Degenerative changes revealed by radiographs of the knee may certainly account for pain. However, it is common for me to see severe degenerative changes on x-rays in joints that are completely pain free! Therefore, I understand the pain as an interaction somewhere along the spectrum between the physical and the psychosomatic. In other words, the osteoarthritis that shows up on the x-ray may trigger the psychosomatic process, which could transform an otherwise asymptomatic or minimally painful joint into a painful one or intensifying the pain beyond what would be expected for that condition.

Further evidence that suggests a psychosomatic component to knee osteoarthritis is the timing over which people often develop pain.

A fifty-five-year-old woman came to my office complaining of a right knee pain she awoke with approximately one month earlier. She had recently convalesced from low back surgery for sciatica, and upon becoming more active, her knee became extremely painful. Prior to her back surgery, she underwent a total left knee replacement due to osteoarthritis. Her past history also included bilateral carpal tunnel surgery, a total hip replacement due to osteoarthritis, plantar fasciitis, obesity, chronic obstructive pulmonary disease from heavy smoking, and depression. Radiographs in the office revealed moderate degenerative changes consistent with osteoarthritis.

This patient has had multiple surgeries to "correct" orthopedic problems. When one problem is resolved, another one develops—an example of the symptom imperative or the ongoing equivalency response. Her x-ray changes had been present for some time, yet her knee had been pain free until she recovered from her back surgery. This is a very common pattern I see daily in my practice, which suggests that many of the common disorders that are thought of as structural in origin have a significant psychosomatic component. Again, I think of these disorders as a spectrum. For some patients, their pain is mostly psychosomatic; for others, there is less of a psychogenic component.

In another case a seventy-six-year-old man came to me with moderate right knee pain and swelling after going up and down a ladder putting up Christmas lights. In the past he had occasional aching in his knee if he "overdid it." Physical examination revealed a small effusion of the knee, decreased range of motion, and pain with flexion. Radiographs revealed severe osteoarthritis of both knees. The patient received a cortisone shot in his knee, and his pain resolved, except for the occasional intermittent aching he described earlier.

In contrast to the previous patient with a painful knee, this patient experiences only intermittent pain from severe degenerative changes depending on his activity level. He promptly responded to treatment measures when he had a flare-up induced by a marked increase in loading of the knee. I believe this patient's knee pain was closer to the structural end of the spectrum.

THOUGHTS ON TREATMENT OF PSYCHOSOMATIC DISORDERS

Acceptance and understanding are the foundation for successful treatment of a psychosomatic disorder. When patients request an appointment with me to treat their musculoskeletal disorder through the methods outlined by Dr. Sarno, they are first screened by a phone conversation to determine their level of acceptance and ability to understand the basic principles of a psychosomatic disorder. Their one-hour visit with me consists of the following:

1. Confirming that the disorder is in fact psychosomatic by taking a history and doing a physical examination and reviewing pertinent imaging studies

2. Carefully assessing their acceptance and understanding of their disorder

3. Reviewing the treatment principles

In the case cited earlier of the forty-two-year-old woman with chronic neck pain, her treatment success was facilitated by several important points emphasized during our visit. First, confirmation that her neck pain was indeed psychosomatic eliminated her fear of the pain and gave her the confidence to resume the activities that she avoided. In other words, she could begin to deprogram herself. Second, she needed reassurance that having painful unconscious emotions is normal, that we all have them, and that she didn't necessarily need psychotherapy to eliminate her pain. Finally, I reviewed with her the important concept that pain serves to distract the mind from the emotional to the physical. We devised methods that reinforced the need to bring back the pain to the psychological, which will eventually eliminate the *distraction* of physical pain.

I have observed that there are several common questions patients express with regard to the treatment of psychosomatic disorders. Foremost is the question of how to "figure out" their stresses and make the necessary changes. This question reflects their understanding that life is too stressful and therefore they must eliminate stress in order for the pain to go away. Of course, this thought process is not correct. First, psychosomatic disorders arise from the unconscious mind, which is where emotions reside that, for the most part, we are unaware of. Not only is it not possible to directly access these emotions, but it is not necessary for successful treatment for most individuals. Successful treatment requires one to simply acknowledge that these painful, unpleasant, often threatening emotions exist. It is not necessary to "figure them out." Treatment is about acknowledging their existence, not changing them. Stress is unavoidable and a part of life. Acceptance, then, not only means accepting one's pain as psychosomatic, but also coming to terms with our genuine self, both the parts we like about ourselves and the parts we don't like. Accepting our painful unconscious emotions as part of who we are is not only a step toward successful treatment but a step toward being a more whole human being.

Another common question patients ask me is whether they should stop their rehabilitation exercises, or going to the chiropractor, or their lumbar support, and so on. The answer is yes! The cause of a psychosomatic disorder, and therefore the cure, resides in the mind. Patients become *programmed* very quickly by well-intentioned advice from various health care providers. For example, when a patient with low back pain is undergoing physical therapy and the therapist advises good posture, proper lifting techniques, and avoidance of potential aggravating activities, the patient becomes programmed to potentially have pain if the advice is not fulfilled. Who has good posture all the time? Successful treatment, therefore, includes identifying the ways in which the patient has been programmed and to confront and overcome them. I have found that treatment failures often stem from the continuation of the intense programming that occurs as an inevitable part of their prior treatments. Tied into this programming is what I call the *fear factor*. Patients become quite fearful that if they do something wrong, such as bending over without keeping their back straight, they will have pain or do more damage. Of course, the pain from a psychosomatic disorder is harmless (not to minimize the intensity of the pain). Once patients accept this, they can confront the ways that they are programmed and overcome them.

FINAL THOUGHTS

My understanding of psychosomatic disorders and my ability to diagnose them is a process in evolution. Is any medical disorder ever completely separated from the mind? What role do emotions play in autoimmune disorders such as rheumatoid arthritis or multiple sclerosis? In cancer? All human bodies fail at some point regardless of their emotional health. What role do emotions play in someone whose body is wearing out or in the process of dying?

Returning to the cases described at the beginning of the chapter, all but the last case are daily occurrences in my nonoperative orthopedic practice. They typify the universal nature of psychosomatic disorders and the common occurrence of the symptom imperative, or the need to keep those painful unconscious emotions repressed.

Edward Shorter titled his book, which chronicles the history of psychosomatic disorders, *From Paralysis to Fatigue*. I would rename the book *From Paralysis to Fatigue or Chronic Pain*. Chronic pain disorders are currently epidemic. However, it is rare for me to mention to a patient the possible psychosomatic causes of their pain. Most patients who come to me to treat their pain as a psychosomatic disorder, such as the last case example at the beginning of the chapter, have almost invariably experienced failure with conventional medical treatments, are desperate, and had a friend tell them about Dr. Sarno's books. Furthermore, these patients need to have the motivation and capacity for insight into their emotional world.

How do I reconcile my dual existence in treating patients with the standard medical approaches while understanding many of their disorders as psychogenic in origin? First and foremost, if I choose to practice in our society, I must accept the realities of medicine as it exists today. And not all pain is psychosomatic. For pain that is predominantly psychogenic in origin, I at least know the true etiology of the pain so I don't get frustrated when it doesn't get better or an equivalency response occurs. I expect it! The way I talk and interact with patients has also changed with this knowledge. I give patients *permission to get better* and don't say things that program them to have pain.

A fifty-two-year-old woman came to me complaining of ankle pain she had had for two years. When she was nineteen she had broken her ankle and required open reduction and internal fixation. The orthopedic surgeon said to her that "you will probably have arthritis when you're fifty." She spent the entire middle years of her life *expecting* to get ankle pain when she was fifty, and sure enough, as soon as she reached that age, she got it!

Many patients with psychosomatic disorders like it when doctors tell them they are damaged goods. It validates their symptom imperative. However, when we doctors do this, we become part of the problem and reinforce the psychosomatic process. I try to present an accurate but hopeful view of their pain. Knowledge of the psychosomatic process has also given me the ability to treat conditions that are thought of as incurable, such as fibromyalgia. Textbooks state that fibromyalgia is of unknown cause and is not curable. I've cured patients with fibromyalgia. They are the patients who are willing to look at their problem as a psychosomatic disorder, like the one I described at the beginning of the chapter. Finally, I hope through my successes and willingness to share my own experiences and the experiences and wisdom of Dr. Sarno I can open the minds of my colleagues and slowly change the way medicine and society view and treat pain and illness.

TEN

A FAMILY DOCTOR'S EXPERIENCE WITH MINDBODY MEDICINE

Marc Sopher, M.D.

Dr. Marc Sopher is a family physician who has practiced in Exeter, New Hampshire, since 1990, with the exception of a year at the University of Vermont, where he worked as team and staff physician. In addition to practice and work with TMS, he also served as medical director of the Synergy Health and Fitness Center and provided medical care to the students of Phillips Exeter Academy.

An avid athlete, Dr. Sopher has run thirteen marathons and the Mt. Washington Road Race three times. As captain of the Williams College tennis team, he led his team to the inaugural NESCAC team championship and continues to play competitive tennis. He enjoys biking and hiking with his family and was once spotted carrying an injured ninety-pound dog on his back down Mt. Washington's Tuckerman's Raving trail, relishing the extra workout.

Dr. Sopher's contribution to this volume is very important since he has applied the principles of mindbody medicine to a family practice. He has seen the role of unconscious phenomena at work in a wide variety of clinical conditions and, like Drs. Hoffman and Rochelle, experienced the frustration of having the mindbody diagnosis rejected by most patients.

THE FAMILY PRACTICE

While Dr. Sarno's practice has focused more on neck, back, and limb pain, I have had the opportunity as a family physician to help many with symptoms encompassing a wide spectrum of psychosomatic disorders. A traditionally trained physician, I have been using Dr. Sarno's approach with great success since reading *Healing Back Pain*, his second book, and eliminating my own low back pain that had plagued me for nearly two years, along with an intermittent sciatica of more than fifteen years' duration. Intrigued that reading a book could cause years of discomfort to vanish, I contacted Dr. Sarno, who invited me to New York to train with him.

After spending time with Dr. Sarno, it was suddenly clear to me that psychosomatic medicine needed to become a more integral part of my practice. It helped to explain why so many of my patients had chronic and recurrent symptoms, despite the utilization of treatment modalities endorsed and encouraged by both traditional mainstream medicine and alternative therapies. I believed that my medical education had been excellent, teaching me when to prescribe medication; when to recommend rest, ice, heat, physical therapy, or special exercises; when to refer to another physician or practitioner for further evaluation; and when to advise injections, surgery, acupuncture, or other treatment modalities. What I realized was that there was almost no mention of the role of the mind as a cause for symptoms. Stress was always mentioned as a factor for certain disorders like headaches, ulcers, and irritable bowel syndrome, but nothing more. Now it became clear to me that virtually all my patients had experienced psychologically induced symptoms sometime in their lives.

In the course of this chapter I shall use the terms *psychosomatic* and *psychogenic-induced* symptoms interchangeably, recognizing, as Dr. Sarno points out in chapter 1, that psychosomatic is one class of psychogenic disorders. Almost all the psychogenic disorders I have treated are psychosomatic.

Prompted by requests from patients with psychosomatic symptoms mentioned but not broadly covered in Dr. Sarno's books, I have written a book, *To Be or Not To Be . . . Pain Free: The Mindbody Syndrome,* based on my work with a wide variety of these mindbody disorders. As a family physician, I care for patients of all ages, from infancy to old age. Family doctors provide comprehensive care of their patients—they treat the whole person. I am responsible for not only evaluating and treating signs and symptoms of illness and disease, but also helping to keep my patients well. I help my patients of all ages to get well when they are ailing and to stay well. Like most family doctors, I am usually the first person my patients seek out to evaluate their symptoms and examine them. The ability to recognize a psychosomatic process when it begins is of great practical value since it spares patients unnecessary and inappropriate treatment, which is usually unsuccessful, and only prolongs their discomfort and disability. Timely, accurate diagnosis and treatment speeds the resolution of symptoms and restores the quality of life, which is our ultimate intention.

Responsibility for the *whole person* leads to encounters with a broad spectrum of mindbody disorders. This primary care perspective can be quite challenging and even daunting at times. While many people come to me because of my reputation for TMS treatment, most of my patients are unaware of that part of my practice. I must introduce them to these new, nontraditional concepts and, based on a variety of factors, they will either accept or reject them. This is in contrast to Dr. Sarno's patients, most of whom are familiar with his ideas and seek him out because of them. Because of the primary care nature of my practice, I see a wide variety of psychosomatic disorders that are in essence equivalents of TMS, in that they stem from the same psychology as TMS. It is my hope that my experience will interest practicing physicians, particularly family physicians, internists, and pediatricians who are rendering primary care.

Not all pain is due to TMS. However, I do believe that the majority of chronic and recurrent pain does not have a structural-physical basis, but a psychological-physical one. On an average day I see between twenty and twenty-five patients. Some visits are well-child visits to update immunizations and review growth and development. Similarly, some visits are for annual adult exams and surveillance for cancer, hypertension, lipid abnormalities, diabetes, tobacco abuse, and so on. Approximately half of the patient visits are for acute illness, such as respiratory, gastrointestinal, and skin infections. But on any given day I will see at least three or four people who wish evaluation of chronic or intermittent symptoms of a broad nature. They may complain of pain in the back, neck, shoulders, arms, wrists, hands, hips, knees, feet, abdomen, genitals—anywhere. They often come in believing they have and should be diagnosed with arthritis, disk disease, tendinitis, bursitis, sciatica, rotator cuff tear, iliotibial band syndrome, migraine headaches, irritable bowel syndrome, dyspepsia, gastroesophageal reflux disorder (GERD), carpal tunnel syndrome (CTS), plantar fasciitis, temporomandibular joint syndrome (TMJ), or fibromyalgia, to name a few. Often, they believe that their work is to blame. Other times, they believe they have an "old injury" that never healed or that they have simply "worn out."

Many have already had many types of treatment, with no response to some and only temporary response to others (placebo). These treatments have included oral medications, injections, massage, surgery, orthotic devices, and so on. They do not have long-term success because the treatments are based on erroneous diagnoses. Treating a psychologically caused symptom with a physical modality like those noted above is bound to fail.

All of this goes back to the concept of differential diagnosis—the process by which the physician obtains information from the patient (the history), does an appropriate physical examination (the physical), orders diagnostic studies (blood tests, x-rays, MRI, etc.), and then interprets this amassed data to produce a list of diagnoses that more or less correlate with the accumulated data. This list is known as the *differential diagnosis.* One of these is chosen as the most likely diagnosis and appropriate treatment is instituted. This is the art and science of clinical medicine—selecting the proper diagnosis and treatment.

One of the major problems in the contemporary diagnostic practices of pain disorders is that physicians *often do not have a differential diagnosis.* With low back and leg pain, for example, if a lumbar spine MRI shows a herniated disc, it is assumed, a priori, that this is the cause of the pain, even when there is a lack of correlation between the location of the herniated disk and the location of the pain and/or examination findings. Those of us who practice psychosomatic medicine are chided for our lack of "scientific evidence" for the existence of such entities as TMS, but neither do the critics have any evidence for their diagnostic conclusions. They speak of compression and inflammation, but there is not one whit of evidence in the medical literature that supports their diagnostic ideas, despite which they blithely proceed with surgical and other treatments. They ignore the lack of correlation between the findings of imaging studies and patient symptoms because they do not have (nor do they wish to have) a differential diagnosis. They are committed in advance to one diagnosis.

My differential diagnosis in this situation would be:

1. The symptoms and findings are due to the herniated disc.

2. The symptoms and findings are due to TMS.

Inappropriate treatment based on erroneous diagnoses is almost the rule where musculoskeletal disorders are concerned. The result is that patients continue to have symptoms, may have temporary relief due to the placebo phenomenon, sometimes develop symptoms in a different location, or even in a different system (gastrointestinal, dermatologic, etc.). This is the symptom imperative referred to earlier in the book.

Our choice of diagnosis is supported by our therapeutic success. In the old days in medicine, successful treatment was considered proof of the accuracy of diagnosis.

Let me be clear. I truly believe that the physicians patients see are in medicine in order to help others. They do wish to heal, to make people well—it is a worthwhile and gratifying endeavor. So, based on their knowledge and experience, they honestly and sincerely offer treatment that they believe will help. *But their knowledge base is not complete.* That last sentence has not won me friends in the medical community, but I have no doubt about its truth. As long as physicians are not aware of the critical role of psychological factors in the causation of physical symptoms, they will be hampered in their efforts to heal.

Why are physicians so reluctant to embrace TMS theory? For one thing, it is difficult to measure. The scientific approach mandates that any treatment be evaluated by formal testing, involving control groups, "blind" evaluations, "double blind" protocols, and the like. Too often, TMS physicians are dismissed by colleagues who state that the TMS treatment results are "anecdotal." The implication is that our results are invalid because we do not employ a scientific protocol. Unfortunately, a control study would be impossible because therapeutic success in the treatment of TMS requires that patients accept the fact that their symptoms have a psychological basis. If patients cannot repudiate the structural explanation for their pain (a disk problem, a heel spur, carpal tunnel syndrome, etc.) and attribute the pain instead to TMS, they cannot get better. In a control study, patients are assigned randomly to two or more treatment methods. If most of the patients assigned to TMS treatment cannot accept the diagnosis for reasons to be discussed, the study is invalid. The nature of TMS puts us at great disadvantage in such studies.

In addition to being difficult to measure with traditional scientific protocols, it is extremely time consuming to treat TMS. It is much simpler to prescribe a pill or recommend physical therapy or surgery than to explain to someone how his *very real* physical symptoms can have a psychological cause. If this concept is entirely new to patients, they are likely to be very disappointed. More often than not, they were hoping for a "quick fix"—some physical treatment that would quickly alleviate their pain. Based on their conditioning and experience, this is often the expectation. Upon hearing that it is likely that their symptoms have a psychological cause, many believe they are being told their symptoms are not real, that they are imaginary. Worse yet, they may believe they are being told that they are hypochondriacs, that they are "crazy," or that it is "all in their head." This can strain even the best doctor-patient relationship. Much time must be spent carefully explaining how psychology can and does affect physiology. It is so much easier to write a prescription!

Sadly, the majority of patients that I introduce to TMS concepts are not receptive. I would estimate that only 10 to 20 percent are intrigued and enthusiastic. The interested ones are often relieved to hear that I do not believe that they have a serious physical problem or disease. They may even be overjoyed to learn that I am going to recommend only some reading and the use of their brains. This group is happy not to have to take medication or be referred for some other treatment. Why it is that more do not respond this way is beyond me. As a society, we are becoming so passive that we wish to be only recipients of treatment as opposed to active participants in our health and wellness. As for the others, as soon as they hear me say *psychological* or discuss the role of stress, they may tune me out or get angry. They may feel that there is a stigma associated with anything psychological and that I do not really understand them. And one dare not use the term *psychosomatic*!

Sometimes I wonder if the public and mainstream medicine will ever wake up. But I still believe there is hope, and a quick story helps explain my optimism.

Glen came in not long ago for evaluation of low back pain. He had previous episodes of low back pain, sometimes with radiation of pain into one leg or the other, but this bout was particularly severe. In his early forties, Glen was married with two children, one a preteen and the other a teenager. Though he enjoyed his work as a commercial pilot, it had for obvious reasons become increasingly stressful since September 11, 2001. After carefully going over his history and doing an appropriate examination, I told him that I thought he had TMS and spent a good deal of time explaining this. He did not take my diagnosis at all well. Incensed, he told me that he had "real pain" and stormed out of the office.

Two days later, he showed up in my office without an appointment and asked to speak with me. He told me that after he angrily left my office, he calmed down and reminded himself that I had seen him and his family through some significant illnesses and difficult times. I had told him that there were no side effects to reading, so Glen reviewed the material on my website, www.themindbodysyndrome.com, and read *Healing Back Pain* by Dr. Sarno. He saw himself in those pages, and when he did, his pain simply vanished. Needless to say, I was surprised and delighted that he had come to apologize and tell me that I was right!

INTRODUCING PATIENTS TO TMS

In my own book, I start off with these comments: "You are probably in pain right now. That is why you are holding this book in your hands, looking for some relief. Perhaps you picked this up because you have heard of Dr. Sarno and TMS. Maybe a friend recommended this to you or you simply discovered it in the process of searching for answers. Your pain may be in your neck, back, legs, feet, head—it could be anywhere. With the information in this book, I am optimistic that you will be able to eliminate your pain, no matter where it is. You will do this with *knowledge*. Simply by changing how you think about the connection between your brain and body, you will begin to feel better. I will not be recommending oral medication, special exercises, surgery, injections, physical therapy, manipulation, acupuncture, massage therapy, prolotherapy, or any other of the multitude of alternative therapies that have sprung up in an effort to combat the explosion of chronic and recurrent pain in our society. Just knowledge.

"Through the process of education, you will gain a better understanding of how *psychology can affect physiology*—how your brain can be responsible for the creation of very real physical pain. Armed with that knowledge, you will do battle with your brain and stop the pain."

When I am explaining the TMS approach to patients, I provide a similar introduction. It is absolutely essential that they understand that you believe their pain is real, that you are not minimizing or brushing off their complaints. Validation of their discomfort is critical to the success of the doctor-patient relationship. I believe it is even more important with treatment of TMS. Only with complete trust and confidence can we expect someone to effect a paradigm shift in their thinking about the mindbody connection.

I then embark on a discussion that I refer to as Psychology 101.

We are sentient beings. We have the capacity for thought and emotions. This is what makes us capable of the most extraordinary achievements—works of art, scientific discoveries, literature, technology, and so on. It is also our downfall. Thinking and feeling allow us to experience both positive and negative emotions. We all seek joy and happiness, but reality intercedes and we all experience sadness and disappointment, anger and frustration. The ability to comprehend the concept of a future carries with it the somewhat less charming sensation of worrying about that future.

As I said earlier, life is stressful. Even if we are happy and feel good about our families, jobs, and finances, we all experience stress. Stress, anger, and conflict arise from three main sources. There are, first of all, the everyday issues such as our home and work responsibilities, worry about our children, worrying about our parents, inconsiderate drivers, the long line at the market, and the like. Second, some of us have experienced much emotional distress in childhood. Even if we have made peace with it, that distress is still there, a potential source of unpleasant feelings. Third, our own personalities also predispose us to these troubling emotions. If we have high expectations for ourselves, if we are ambitious and place great demands on ourselves, if we are very conscientious about our performance, then these perfectionist traits are causes of stress. If we go out of our way to help and care for others, even to the point of self-sacrifice, then these "goodist" traits also create stress as we make our needs subordinate to those around us.

These personality traits are not undesirable—they make us successful, kind, and considerate. But it is essential to understand how these very qualities can contribute to the accumulation of stress, anger, and conflict. The way our brains work, we repress unpleasant thoughts and emotions, which then find a home in the unconscious. This is a very good defense mechanism—it allows us to move on and take care of our responsibilities and be nice people that others like and respect. Unfortunately, we can only hold so many of these unpleasant thoughts and emotions in the unconscious. Accumulated anger, stress, and conflict become *rage*. This rage wants to rise to consciousness, but we usually do not let this happen. If it were to happen, we might rant and rave and do things that would not be acceptable, things that would make others not think well of us. To distract us from these unpleasant thoughts and emotions, our brain creates pain, real physical pain. In our society it is acceptable, even "in vogue," to have certain symptoms, such as back pain, headaches, and acid reflux. When we focus on our pain, we are distracted from these causes of *rage*. This is a brilliant strategy on the part of the brain. Why does this occur? No one can know for sure, but we know this happens because by learning about it, we can stop it. We can stop it and thereby eliminate the pain.

So, the unconscious mind is the site of repressed and suppressed emotions. It is where the reservoir of rage lurks. The *reservoir of rage* is Dr. Sarno's term, and I think it provides a compelling image for the origins of pain.

To summarize: Dr. Sarno has identified three potential sources for this rage in the unconscious. In each person the quantity from each source will vary.

1. Stresses and strains of daily life

2. The residue of anger from infancy and childhood

3. Internal conflict (self-imposed pressure—the clash of the id and the superego; it also comes from perfectionist and goodist traits)

Unpleasant thoughts and emotions may be pushed into the unconscious, as they would be difficult to bear. If we attempted to deal with them, it is possible that we would somehow become incapacitated in one of two ways. First, the id could take over, in which case angry, belligerent behavior would occur. In my lectures I refer to a ranting, raving lunatic, someone in need of a straitjacket. But behaving like that is not acceptable, so we push those thoughts away rather than act inappropriately and be ostracized (causing further reduction in self-esteem). Second, we could become paralyzed with grief, unable to function in the face of unpleasantness. But we don't want that either, because then we'd be shirking our responsibilities.

Repressing (unconsciously) or suppressing (consciously) thoughts and emotions that are unpleasant, disagreeable, or unacceptable allows us to carry on, but adds to the reservoir of rage. It helps to think of rage as accumulated stress. Not all sources of stress are equal—

some may be annoying nuisances, while others may be enormous. This is a critical concept. I have seen many patients who struggle with it. If they are unable to conceive of a source of rage or a serious stressor, they may doubt that they have such a reservoir in their unconscious. But it's important to remember, the reservoir can fill with unpleasant thoughts and emotions of all sizes. And another very important concept: reservoirs come in all sizes.

A common misconception is that the onset of pain must coincide with some obvious source of stress. While this can sometimes occur, like getting a headache on a bad day, it often is not the case! This can be a difficult obstacle for people to get over. Many times, people insist that everything is *fine,* that the pain began on vacation or when everything in their life was grand, that they didn't do *anything* to bring it on. They will say, "Why now?" This may cause serious doubt for them that TMS could be the cause. Go back to the reservoir of rage. There is always stress, even if life is good! We all worry to some degree and we all have the eternal, internal conflict between the id and superego. Like the straw that breaks the camel's back, some little unpleasant thought, emotion, or stressor is tossed into the reservoir, which is now threatening to overflow. The brain will not allow it to overflow or rise to consciousness—and in order to distract us and keep the reservoir and its contents hidden in the unconscious, it creates pain. And perhaps, just perhaps, by creating pain, the brain not only causes distraction but the expansion of that reservoir.

Now you understand about the reservoir of rage. These unpleasant thoughts and emotions "strive to rise to consciousness," which would be completely unacceptable. To prevent this from happening, the brain creates pain as a distraction. As a society, we are very somatically focused, preoccupied with every ache or pain. By focusing our attention on physical symptoms, we keep these painful thoughts and emotions repressed. This is a very effective strategy, and as evidence of that fact there is an absolute epidemic of mindbody disorders in our society.

Because full understanding of these points is crucial for symptom resolution, I then reiterate: *TMS is a strategy of the brain to keep unpleasant thoughts and emotions from rising from the unconscious into the conscious mind. The brain, through established physiologic pathways, creates pain as a distraction. By focusing our attention on physical symptoms, we keep these painful thoughts and emotions repressed. This is a very effective strategy, as there is an absolute epidemic of mindbody disorders in our society.*

I then review some examples, including their own symptoms, and explain:

Eliminating the pain is startlingly simple. We can banish the pain and thwart the brain's strategy by simply understanding and accepting that the pain has a psychological causation, that it is not physically based.

This is a simple concept, but sometimes challenging to implement. When I first met Dr. Sarno, I had a number of questions about TMS treatment and how to help my patients. My number one question was: is it really as simple as getting patients to learn how to think differently? The answer is yes, but with the understanding that the mental work involved may be considerable. It entails *forgetting everything you've ever been told about your body and symptoms.* This can be very difficult for some. I refer to this process as undoing old ways of thinking, and then move on to a discussion of the principles of conditioning. We are conditioned to hold certain beliefs about ourselves, as a result of the explanations and comments of well-meaning health care providers, family, friends, neighbors, coworkers, and the media. Repeatedly, we hear that physical symptoms must have a physical cause, that we are inherently fragile and susceptible to injury, that certain "injuries" can result in chronic pain, that healing can take a very long time, that we have to learn to live with certain kinds of pain. If we believe in these assumptions, which I refer to as *myths* in my book, then we come to expect and accept certain kinds of pain as a result of our activities. Thinking differently requires reprogramming the mind, enacting new modes of thought—new conditioning to replace the old conditioning.

I think this helps to explain one of the more curious aspects of TMS treatment:

Why do some people, who agree they have TMS, get better more quickly, others more slowly?

And in the same vein:

Why do some people feel better after just reading my book or one of Dr. Sarno's books?

I have puzzled over this and concluded that the "rapid" healers are somehow better able to put aside what they have been told in the past and fully integrate the TMS information. They can undo the conditioning that is part of mainstream thought and replace it with this new understanding of how the workings of the unconscious can affect the body and physical sensation.

STRATEGIES OF TMS TREATMENT

Make a List

Think of anything that could be a source of stress for you. Think about what makes you angry or enraged. Think of what things you worry about. Think about your personality. Identify perfectionist and/or goodist traits. Are there people in your life who did not treat you as well as you would have liked? Write all of this down. It is impossible to know what is in your unconscious (hence the name *unconscious*), but it is possible to contemplate what might be there. By acknowledging the presence of these unpleasant thoughts and emotions, you can thwart the brain's strategy. As you undoubtedly recall, the brain's strategy is to create pain, pain that will serve as a distraction. Focusing on the pain is a type of defense mechanism—it keeps us from thinking about those things that make us upset, worried, or angry. The pain keeps the reservoir of rage hidden. When we recognize that it is there and what it may contain, there is no need for the pain, because there is no further need for distraction.

Making a list is like keeping a journal. Many studies have shown that those who write regularly in a journal, about themselves, their thoughts and concerns, are healthier than those who do not. So, start your list or a journal, and add to it or review it regularly.

Reflect

By now, you have figured out that it is the process of self-education that will help you to feel better. It is amazing—no medication, no physical remedies, and no side effects. Set aside time each day to think about TMS theory and treatment. Read and reread my book and Dr. Sarno's books. It's not necessary to reread everything, but it will be helpful to reread passages that you find particularly pertinent. Even when you feel well, spend some time each day on this material. This will help you to remain well. It is good preventive medicine.

Discard Your Physical Remedies

Get rid of the special back supports, heel pads, orthotics, pillows, chair cushions, and so on. They cannot fix the problem, and you don't need them. Physical modalities cannot help symptoms that have a psychological cause. Their very existence is part of the old conditioning and will only perpetuate the symptoms.

If you are taking narcotic pain medications, you will need to wean yourself off these gradually under a physician's supervision. Similarly, you should also wean yourself off of benzodiazepines (such as Klonopin, Ativan, Valium, Xanax, etc.). These medications only mask symptoms and cannot cure them. In addition, they are physically and psychologically addicting and will only perpetuate the symptoms. They will also impair cognition and interfere with your efforts at self-education.

It is reasonable to take non-narcotic medication for pain, such as aspirin, acetaminophen, ibuprofen or naproxen (all available over the counter). However, each time you do, it is important to remind yourself that these drugs will not fix the cause of the symptoms and will just temporarily take the edge off while you continue to apply yourself mentally.

There are a myriad of other medications prescribed for the host of ailments discussed here. In most cases, medication can be safely discontinued, but this should always be discussed with your physician first.

Be Eternally Vigilant

Eternal vigilance is the proverbial "ounce of prevention." This is why it is necessary to spend some time each day reflecting. Celebrate the good days. This is essential to reversing the old condition. Tell yourself you are indeed fine—if you had a physical problem, where did it go? However, do not be discouraged if pain returns or occurs at another location. Remember, your brain does not want to give up its strategy of distracting you with pain—this is how we are made.

Resume Activity

You are not really well until you are back doing the activities you formerly enjoyed. While you may have to start slowly (it is still necessary to follow appropriate guidelines for exercise training), you should be able to do whatever you want. We are capable of far more than we have been told. I think very few of us approach our potential because we have been misinformed about the limits of our bodies. I have patients in their sixties, seventies, and eighties running marathons, bicycling across the country, climbing mountains, and participating in other strenuous activities. They are not supermen and superwomen; they are simply folks who have taken good care of themselves and refuse to believe that they are fragile.

Another point that is integral to reprogramming your mind bears mentioning. Attacking TMS symptoms *does not require positive thinking*. While it is good to think positively and have an optimistic outlook with regard to yourself and life in general, it is not positive thinking that will cause symptom resolution. If it were so, most of you would not be reading this, and there would not be an epidemic of mindbody disorders. How do I know this? Virtually everyone *wants* to be well. It is only the rare individual who wishes to experience pain and suffering. Most people try very hard to ignore their symptoms and to soldier on. They try to think positively; they try to put "mind over matter." In one form or another, this is what most self-help books promote. "Think positively," "just do it," and "mind over matter" are common themes. Others focus on stress management, behavioral modification, and relaxation techniques. Don't get me wrong—these are great skills to have. Undoubtedly, we could all do better with stress management and could benefit from honing these skills. However, this is not what will eliminate pain. It doesn't require positive thinking. It requires *accurate thinking*. Accurate thinking means understanding how psychological factors affect our physiology. Only when we understand this can we truly heal ourselves.

It's not easy forgetting all that you have been told and, in essence, creating a new belief system. As a matter of fact, it is extremely difficult. There are many obstacles, both within and without. Many people speak to me about fear. Invariably, each has undergone a comprehensive evaluation by their physician (or multiple physicians). They may have been told they have one of the diagnoses that I have mentioned here. Very possibly, they have been told that they must avoid certain activities or they will risk further damage or escalation of symptoms. For many, this can be devastating, particularly if they have been advised to give up or curtail an activity that has brought them much pleasure. I have dealt with runners, cyclists, tennis players, hikers, and the like, who were despondent about giving up or reducing their form of exercise. Even when they say they believe TMS is their problem and I've told them to resume exercise, they admit to being fearful that their symptoms will recur or increase. Fear is powerful, and it is part of the conditioning that has occurred over time. It takes courage to put aside the fear.

Even when someone tells me they have returned to their previous activity with minimal or no pain, they may admit that they remain nervous or fearful about the next time. In many cases this may be a reflection of their personality, as well as their previous conditioning that needs to be undone. Remember, many people with TMS are prone to worrying. They may be perfectionists, placing much pressure on themselves to do well, or to succeed or be well thought of, or they may be concerned about their ability to be of service to others. They may also have a more simple fear that their symptoms represent a physical decline or deterioration that heralds future morbidity or mortality.

So, whenever someone confronts their fear, resumes their activity, and feels fine, I tell them to celebrate. CELEBRATE! I tell them to talk to their brain—tell themselves that they are fine! There cannot be a physical problem if they were able to do the activity without difficulty. *Celebrating is an important way to reprogram the mind.* It helps condition you to think differently about your body and will help you immeasurably to undo the old conditioning and forget all that came before.

On the flip side, it is important not to be discouraged if symptoms arise during the course of an activity. It simply means that more mental work must be done. It is easy for fear and its gloomy companion doubt to creep in. "Maybe it isn't TMS, maybe I do have a physical problem" is a common thought. The best advice is to simply acknowledge this fear as part of the old conditioning, to recognize that it is simply an aspect of the brain's strategy to have you believe there is a physical problem.

A common question I hear daily is "What should I do when I have pain, especially a lot of pain?" It can be very difficult to ignore pain and carry on. First, you must talk to your brain and remind yourself that you are physically fine. Tell your brain that you are onto its game, that you know about the reservoir of rage. Like Dorothy discovering the Wizard of Oz behind the curtain, you won't be fooled! The pain is not present because you've done something that you are incapable of or that you are so feeble or fragile. Try to pay it as little attention as possible. Remember, the brain wants to distract you and keep your attention and focus on pain, rather than on what may be in the unconscious. Many patients become obsessed with their pain—they must learn to shift their focus (this is the reprogramming, or reconditioning process). Try not to give in! Try to remain active, doing the activities that you enjoy.

If someone states they truly believe that TMS is the problem, that they have been doing the mental homework and yet are distressed that their symptoms persist, they may question whether they have TMS. This has the elements of a Catch-22. If you begin to doubt there is a psychological cause, that there could be a physical cause, then the work is undone and the brain's strategy of creating a physical distraction will triumph. This is part of what I refer to as the *calendar phenomenon:* preoccupation with the number of days or weeks it will take to get better. By this time, everyone may know of someone whose symptoms vanished immediately after reading the book or shortly after seeing a physician trained in TMS treatment. So, an expectation is created in their mind that their symptoms should recede soon after incorporating this philosophy. They look at the calendar and become upset as days and weeks go by. This is where I tell people to look back at their personalities. The calendar phenomenon is another manifestation of perfectionist tendencies—it is self-imposed pressure to succeed and succeed quickly. If they can recognize this aspect of their personality and add it to their "list" of sources of stress, relief will be on the way.

Fear, doubt, the calendar phenomenon, and the failure to think accurately are examples of some of the internal obstacles to healing. Several external obstacles also bear mention:

1. You have read Dr. Sarno's book or mine and become convinced that this approach makes sense. When you mention it to your physician, he/she either dismisses it out of hand or nods indulgently, and advises a traditional regimen, including medication, physical therapy, etc.

2. You have become convinced that this approach makes sense, but when you mention it to your friends, family, and/or coworkers, they look at you as if you have lost your

mind. They, too, may nod indulgently and then recommend a physician, practitioner, medication, herbs, etc.

3. You have become convinced that this approach makes sense, but then you pick up a magazine and read an article discussing symptoms like yours, and there is no mention of TMS as a possible cause. Or maybe, just maybe, there is a brief mention of Dr. Sarno's work with TMS, but other quoted sources dismiss it out of hand. As you trust the members of the media to do their homework and provide accurate, complete information, you begin to wonder whether TMS is for real.

These scenarios occur every day. They may contribute to the conditioning that allows the pain to persist. Even in my own office, when I am introducing one of my established patients to TMS concepts, they may get angry or look at me as if I had two heads. You see, they have come in unsuspecting. They have come in to see me for evaluation of some physical symptom and did not expect to hear that it may have a psychological cause. Some are delighted, enthusiastic, and quite willing to think outside the box. To the others I explain that I can only expose them to this different way of thinking, that I cannot make them believe it. I will certainly try to make my case and be convincing, but it is ultimately up to them to decide.

Perhaps when TMS theory and treatment are finally embraced by the medical mainstream, more people will be open to this way of thinking about themselves. For those who have already reached that stage, it is extremely gratifying to see them succeed at getting rid of their pain and improving their quality of life. Trite as it sounds, I became a physician to help others, to help them when they are ill and keep them well. I am saddened when people refuse to accept the possibility of a psychological cause and to continue to suffer.

CASE HISTORIES

In this section I offer a number of illustrative case histories. Some of these are mentioned in my book but shared again, as I feel their situations will strike a chord with many.

Ken is a forty-eight-year-old man who for more than twenty-five years suffered from low back pain that could radiate down the leg to his foot. His initial symptoms were treated with back surgery—lumbar laminectomy. Never completely relieved, his pain intensified, and he was again diagnosed with a herniated disc. Another back surgery followed, also with incomplete resolution of symptoms. In the year before he came to see me, he had pain with sitting and all activities that he formerly enjoyed, such as bicycling, in-line skating, and hiking. When working at his desk or computer, he would stand instead of sit. He bought a special mattress, obtained orthotics for his shoes, and did special exercises, all in addition to the usual treatment. His most recent MRI, done to evaluate back pain radiating down the leg, was interpreted as showing "scar tissue pressing on nerves."

During our session. Ken described himself as a perfectionist, an overachiever, and a "people pleaser." Though happily married, he identified stress at home with his stepson's learning disability and his widower father living with him. He also recognized that his father had been very critical and emotionally abusive when Ken was younger. Within one month he was much improved, and by three months was virtually pain free and back to enjoying long distance bicycling and hiking. Four years later, he continues to be fine, sending me e-mails chronicling his athletic exploits. It is worth noting that his other TMS equivalents, eczema and frequent urination, also resolved.

Connie described a lifelong history of sciatica. Fifty years old and single, she had leg pain with sitting and running, an activity that she loved. She had given up running at the orthopedist's suggestion after her spine MRI revealed degenerative changes, multiple herniated discs, and scoliosis—in her words, "a mess." Her history also included chronic foot pain, attributed to a Morton's neuroma, which was exacerbated by running.

Physical therapy, manipulation, nonsteroidal anti-inflammatory drugs (NSAIDs), narcotics, benzodiazepines, and eventually surgery all failed to relieve her pain. So unbearable was her discomfort that she admitted to suicidal ideation.

Raised by alcoholic parents and alcoholic herself, she had been sober for a number of years. She had no problem with the idea of a reservoir of rage in the unconscious. Within three months she was much better, back to running and training for a marathon, which she successfully completed!

Paul Teta, fifty-three years old, is another long-term sufferer. Paul's symptoms began more than twenty years earlier while he was playing basketball. Excruciating pain would travel down his leg, and he underwent back surgery for a herniated disc. His symptoms improved for a while but then returned. Sometimes they were severe enough to make it impossible for him to work or do the athletic activities that he enjoyed.

A repeat MRI showed disc herniation, and NSAIDs and narcotics did not ease his pain. Married with two children, Paul owns and operates an auto repair shop. He admits that he is a perfectionist, sometimes "high strung" or "uptight." Not wishing another surgery and wanting to resume his life, he came to see me. Two weeks later, he was fine, and several years later remains pain free. Following is his letter:

> Dr. Sopher—
>
> My name is Paul Teta. I recently (two weeks ago) had an appointment with you. I came with my brother, whom you were nice enough to invite to the seminar. I just wanted to give you a quick update on my progress. The day I arrived for my initial exam I was in pain and also on a strong painkiller, and had been for weeks.
>
> I had read Dr. Sarno's book twice. After you confirmed that I had TMS, you said to not to fear the pain for it was harmless and my back was normal. I think that statement saved me weeks of time. That evening we went to the seminar, which gave me even more confidence. I have not taken any medicine of any kind for back pain. Several days after, I put on my roller blades and bladed about ten miles. At that point my leg was killing me. I continued to blade for another eight miles and my back started to twist and I was losing the lumbar curve. I kept repeating to myself, "The pain is harmless and my back is normal." At about nineteen miles the pain stopped and my leg turned warm. Afterward, I had a twenty-mile drive home. I had no pain sitting for the first time. While driving home I started to scream out loud, "I'm sick of this pain dictating my life!" I began to cry and did so for about thirty minutes (possibly childhood rage?).
>
> After that I have taken out my running shoes (after fifteen years) and resumed running. "NO PAIN." I feel eighteen again. I also can bend over and can for the first time put on my socks without lying down. In my wildest dreams, I never expected to do this well. Twenty years of fear and pain erased so quickly. I have since purchased about ten books and have given them to friends and customers of mine. THANK YOU, DR. SOPHER AND DR.

SARNO. I will send you a future letter of more details as soon as time permits.

Paul and I have kept in contact and he continues to be well, enjoying an active life. He has also helped countless others by introducing his friends and customers to TMS.

Stan is a fifty-three year old whose low back pain began ten years earlier while he was doing plumbing work. His pain would radiate down one leg and his subsequent diagnosis for back pain with sciatica was a herniated disc and degenerative disc disease at multiple levels. He saw an orthopedist, a physiatrist, and a neurosurgeon. Epidural steroid injections, NSAIDs, acupuncture, and physical therapy did not help. Desperate, he went to a chiropractor, who manipulated him for thirty consecutive days. Sitting caused pain. He stopped running, as this also aggravated his symptoms.

Describing himself as "very responsible, to a fault," he put a tremendous amount of pressure on himself. His first marriage, which ended in divorce, had been very stressful. He is now happily married with four children. Work was very demanding. Recently, his mother had passed away, and he was trying hard to improve a relationship with his father that had been poor in the past.

One month later, he was much improved and described being pain free at four months. A couple of years later, he is still well.

Carla, a forty-six-year-old homemaker with one son, complained of neck, back, leg, and foot pain for more than two years. Often, the lower leg pain and foot pain was perceived as a severe "burning" that prevented her from walking. She was evaluated by her primary care physician, as well as an orthopedist, a neurologist, a physiatrist, and a neurosurgeon. Along the way she also developed jaw pain, for which she saw her dentist and then an otorhinolaryngologist. Spine MRI showed disc herniations in her neck and low back. Other tests, including blood work and nerve studies, were normal. She received many diagnoses, including neuropathy, and was put on Neurontin (an antiseizure medicine) when all other treatments failed. She had been active, enjoying bicycling, hiking, canoeing, and cross-country skiing, but had to stop due to pain.

A worrier, Carla noted she was very concerned about her son's safety when he went to college. Her mother had died when she was only two years old, and she was raised by her father's sisters, as her father never remarried. He, too, had passed away recently and had always been difficult, even when he spent his last years in her home. Within two months of her appointment with me her symptoms had resolved.

Larry, thirty-one years old and married for the second time, had been diagnosed with bilateral carpal tunnel syndrome and lateral epicondylitis. He had elbow, forearm, wrist, and hand pain of two years' duration that had failed to respond to rest, NSAIDs, wrist splints, and forearm bands.

He admitted that his symptoms began with an increase in marital stress with his second wife. With children from both marriages, he acknowledged significant financial pressures as well as a desire to be a good father. His pain left shortly after our visit, and he remains pain free more than two years later.

Bill had elbow pain for more than one year when he came to me. He had seen an orthopedist and been treated with physical therapy, NSAIDs, and cortisone injections for lateral epicondylitis—all to no avail.

He had no problem identifying himself as a perfectionist preoccupied with his responsibilities as a husband and father. The cycles of his business were also a source of great stress.

Bill was able to eliminate his elbow symptoms quickly. When shoulder and chest pains then appeared to replace the elbow pain, appropriate studies were done. When all tests came back negative, he agreed TMS was again the culprit and has felt fine since.

Frank, married with two daughters, is in his forties and had frequent episodes of chest pain. His first episode was associated with a viral infection of his heart, known as myocarditis. He had a complete recovery, but for the next four or five years he experienced chest pain that was similar in nature to that which he experienced when diagnosed with myocarditis. He described a constant ache that sometimes lasted for days at a time and worried him greatly. Thorough cardiac evaluations were done several times, all with normal findings.

He admitted that he is "somewhat anxious" by nature. Devoted to his family, he also works two jobs. Always upbeat, he initially thought it unlikely that stress could be to blame for his chest pain, but in time he warmed to the concept. Today, he has been free of chest pain for more than two years. He told me with a smile that the pain had relocated to his elbow, but was now resolving with his use of knowledge.

Jack was a former athlete, now in his forties, with left hip pain. His orthopedist told him that he would benefit from a new hip joint, as his x-ray showed "significant" degenerative changes. After this recommendation, his left hip pain increased and he mentioned it to me at the time of his annual physical exam. When he told me that his right hip felt fine, I asked him to humor me by having *both* of his hips x-rayed. On x-ray, both hips had the same "degenerative" changes, yet his right hip did not hurt! I advised him to put off surgery, resume activity (he had significantly decreased his exercise after being told of his arthritic condition), and not pay too much attention to his hips. Following these instructions, his discomfort subsided and he successfully resumed exercise and athletics.

Jack was able to acknowledge that he was the quintessential

goodist. A devoted husband and father of two boys, he had also taken responsibility for the care of his ailing, elderly parents. As the oldest of his siblings, he made frequent trips out of state to visit his parents, arrange for additional help for them, and coordinate long-term care plans. When they declined further, he arranged for hospitalization and even brought them into his home for respite care. Through this most difficult period he was also busy with his own business, based out of his home region. He understood that he placed inordinate amounts of pressure on himself to ensure the well-being of his loved ones—classic goodist traits.

Steven was a teenager and a budding running star. In the course of his training he began to experience left shin pain. He had not suddenly increased his mileage or suffered any trauma. Methodical in all things, his footwear, nutrition, and hydration are all appropriate. A podiatrist and an orthopedist recommended rest as treatment for what was presumed to be a stress fracture. When he returned to running, the pain returned. A bone scan was ordered and interpreted as showing a stress fracture. More rest was advised. At this point he came to see me. He admitted to being a perfectionist and putting much pressure on himself. Not surprisingly, he was a straight-A student and participated in a host of extracurricular activities in addition to running. After I explained why I thought his leg pain was psychologically induced and running should not cause him pain, he went home and read *The Mindbody Prescription.* The very next afternoon he phoned, obviously very excited. He had just returned from a long run and felt fine! He went on to have an outstanding season, continually lowering his times and improving his performance. His only frustration was an inability to convince his teammates to think psychological and better deal with their "injuries."

Martha, a lovely woman in her late twenties, came to see me for evaluation of chronic lateral knee pain. She was exasperated, as her symptoms had forced her to stop running, an activity that brought her great enjoyment. In addition to her family physician, she had seen an orthopedist, physical therapist, and podiatrist. Rest, therapy, medication, orthotics, and change in footwear and stretching regimen had failed to ameliorate what had been diagnosed as iliotibial band syndrome, a common diagnosis, particularly in runners.

She admitted to being a perfectionist, always putting pressure on herself to achieve. She also admitted to a history of an eating disorder and understood that this reflected issues regarding self-esteem. Single, she was living with her longtime boyfriend, trying to decide whether marriage was in the future.

Soon after her initial visit with me, she resumed running, pain free. She eventually trained for and completed her first marathon.

Barbara came in for evaluation of chronic hip pain. She also noted intermittent heel, knee, and low back pain. Symptoms appeared to have begun around the time of her mother's illness and death several years prior. She worked full time in addition to her responsibilities at home to her husband and teenage children.

She admitted to self-esteem issues and was candid about growing up in an environment with multiple alcoholic family members. Her pain vanished and has not returned since she learned that it was psychologically caused.

Jack is a forty-five-year-old with heel and foot pain for more than one year. Diagnosed by both a podiatrist and orthopedist as having plantar fasciitis, but nothing alleviated his daily foot pain. He tried orthotics, NSAIDs, taping, stretching, and special exercises—all to no avail. In addition to his foot pain, he has a history of chronic intermittent back pain despite two surgeries, reflux, migraines, and irritable bowel syndrome.

Married with two children, he is self-employed and trying to get a book published. He is very happy with his life but acknowledges that he feels much responsibility for his family and realizes that this is a source of stress.

At my urging, he stopped all treatments, and within two weeks his foot pain resolved. His other symptoms also improved.

Renee is a woman in her forties who had been suffering from diffuse myalgias and a multitude of other unpleasant symptoms for more than four years. Diagnosed with fibromyalgia, she described shooting pains in her head and pain in her neck, hips, shoulders, pelvis, arms, and back that could be experienced as sharp, aching, or burning. She also described numbness in her forearms and fingers with keyboard work. For much of this time she also had ear and jaw pain and for the previous six months complained of intermittent nausea, dizziness, and a sense of being off-balance. Along the way she had also been diagnosed with TMJ, carpal tunnel syndrome, irritable bowel syndrome, and reflux. A family physician; a neurologist; an ear, nose, and throat specialist; a dentist; an oral surgeon; and an allergist had evaluated her. Every test, including exhaustive blood tests and imaging studies, came up normal. She had tried everything but the kitchen sink—antidepressant medication, dental appliance, chiropractic, and so on. She had been in counseling for years.

After I explained to her that she had a severe form of TMS, not fibromyalgia, she revealed a personal warehouse of stress that was enormous. Married with four children, she also worked full time as a teacher. Her mother had died suddenly from a cerebral hemorrhage. Her father had died following a long, difficult struggle with multiple sclerosis. One of her sisters had been diagnosed with multiple sclerosis. Another sister was suffering from depression. Within two months of our initial meeting her symptoms were gone and have not returned.

Bonnie is a thirty-three-year-old married woman with severe low back pain that developed after a complicated pregnancy. Pain could travel into either leg, and she also described intermittent pains at other locations, sometimes severe. She was told that her symptoms were due to leg length discrepancy as well as multilevel disc disease, diagnosed on MRI.

When she saw me, she had failed all traditional therapies and was fearful that she would be unable to care for her child or return to work (as she desired). In retrospect, she identified her history of panic attacks, irritable bowel syndrome, and previous episodes of back pain and paresthesias more than ten years prior as earlier manifestations of TMS.

She gave up her lift (meant to treat the leg length discrepancy) and was much better within several months. Three years later, she contacted me to provide an update—not only was she feeling well, but the irritable bowel symptoms that had plagued her for fourteen years were gone too.

Matt, an attorney in his thirties, had previously been diagnosed with chronic epididymitis. He would have frequent episodes of testicular pain. Physical exam and tests were always unrevealing. Antibiotics and NSAIDs did not help. When without testicular symptoms, he often experienced palpitations, tinnitus, elbow pain later diagnosed as epicondylitis, and diffuse gastrointestinal symptoms labeled irritable bowel syndrome.

In going over his history, it became clear that he had easily identified perfectionist and goodist tendencies. He quickly embraced TMS thinking and has been fine since.

Janet had classic symptoms of trigeminal neuralgia for several years when I first saw her. Prior care for this had been with another family physician, a neurologist, and a neurosurgeon. Oral medications prescribed included anticonvulsants (typically used to treat seizure disorders but often used for neuropathic pain syndromes), steroids, NSAIDs, and narcotics. Injections and surgical procedures involving the nerve failed. She was on high doses of narcotics at her first visit with me and acknowledged escalating usage. Like many others, she was at first skeptical on hearing of TMS. Desperate for help, she put aside her reluctance and read Dr. Sarno's *The Mindbody Prescription* and recognized herself on those pages. Married with two small children, she was forced to work outside of the home due to financial pressures. She acknowledged a childhood that was at times very difficult, and she had had little relationship with her father. I helped her to wean herself off the narcotics, and she has remained pain free—and drug free—for more than five years.

Kevin's tale illustrates the unique situation of the elite athlete. He was training for his first Ironman competition when he developed left hip and leg pain just one month before the event. The Ironman is a remarkable endurance event, requiring participants to swim 2.4 miles, bicycle 112 miles (without drafting), and, finally, to run a marathon (26.2 miles).

A veteran of marathons and triathlons, Kevin's pain began when he decreased his training as part of the pre-event taper recommended for this type of competition. Given the incredible amount of training and commitment this required, he was extremely upset about the possibility that he'd be unable to participate. Fortunately, he recognized that his pain could be TMS. Kevin told me that he had successfully eliminated low back pain, shoulder pain, and leg pain, the last diagnosed as iliotibial band syndrome, by reading and rereading Dr. Sarno's books over the past five years. Given how meticulous he was with training and how superbly conditioned he was, he agreed it was extremely unlikely that he should have suddenly developed a purely physical pain problem.

Then, a week before the Ironman, he developed low back and groin pain. I offered him advice on how to conquer his pain and encouraged him to at least start the race. He was in discomfort when he started, but his pain faded and he told me th he was absolutely fine by the end of the race. He said, "Thank you again for your help! I feel extremely fortunate to have been able to finish (he did better than just finish—he achieved an excellent time). It pains me to see so many friends battling TMS-like ailments."

Aaron, an Olympic athlete, contacted me in desperation to explore whether his chronic hip and leg pain could be due to TMS. His symptoms were preventing him from pursuing his living as a professional athlete—sponsors were threatening to withdraw their support. Despite seeing a multitude of physicians and practitioners of all types and religiously following their treatment protocols, he could not run without significant discomfort. Fortunately, one of his family members recommended he read one of Dr. Sarno's books, and an Internet search for TMS physicians led him to me.

During the consultation, he revealed his perfectionist qualities and the extraordinary stress he placed on himself to perform at the highest levels. He was also able to tick off a whole host of significant life stressors, all of which undoubtedly contributed to his reservoir of rage.

He was soon back to running without pain.

Carol contacted me to help her with chronic wrist and forearm pain that had been diagnosed as tenosynovitis and carpal tunnel syndrome. She had been following the treatment plans of an orthopedist and a physical therapist without success. Based on an abnormal MRI, the orthopedist was now recommending surgery.

Since all the physical remedies were failing her, she started to think TMS might be to blame. Carol had managed to eliminate other chronic pain symptoms six years earlier after learning about TMS. What she came to realize this time was that the brain never gives up its strategy—that the appearance of new symptoms, at a new location, is common. What is critical is being able to recognize this. Rather than think, "How did I injure myself?" she reminded herself that she had indeed not done anything extraordinary or unusual—nothing she was incapable of doing. She returned to making her list, thinking about what might be in the unconscious, in the reservoir of rage, threatening to burst into consciousness.

Within one week, she reported feeling "a thousand percent" better! I then reminded her of the need for eternal vigilance. Thinking about TMS, even when we are well, serves as preventive medicine. Those who are able to do this find they can nip in the bud many aches and pains. When this occurs it is extremely empowering, as we recognize the control we have over our bodies. This also helps to solidify the new belief systems about how psychology affects our physical being.

FINAL THOUGHTS

It gives me great satisfaction to be a contributor to this book. Learning about TMS was truly an epiphany for me, and as a result of this education, I have been fortunate to have had great success helping people to feel better. I cannot help but believe that sharing this experience will allow others to experience this same epiphany and resulting satisfaction from improving the lives of patients, friends, and family.

Ultimately, I am confident that TMS theory will become part of mainstream medicine for the simple reason that it is correct and is more successful at alleviating pain than any other modality. As more and more people are helped with this approach, physicians will be forced to take notice. Besides, knowing how awesome and complex the brain is, doesn't it seem rather shortsighted to discount the role that the brain can play with regard to bodily sensations?

REFERENCES

ONE: WHAT IS PSYCHOSOMATIC MEDICINE?

American Psychiatric Association. *Diagnostic and Statistical Manual of Mental Disorders*, 4th ed. Washington, D.C.: American Psychiatric Association, 1994.

Alexander, F. *Psychosomatic Medicine*. New York: W.W. Norton, 1950.

Beecher, H. K. Pain in men wounded in battle. *Annals of Surgery* 123:96–105, 1946.

Beecher, H. K. Surgery as placebo. *JAMA* 176:1102–1107, 1961.

Bengtsson, A., and Bengtsson, M. Regional sympathetic block in primary fibromyalgia. *Pain* 33:161–167, 1988.

Cousins, N. *Anatomy of an Illness*. New York: W.W. Norton, 1979.

Damasio, A. *The Feeling of What Happens*. New York: Harcourt Brace & Co., 1999.

Freud, S. *The Standard Edition of the Complete Psychological Works of Sigmund Freud*, vol. II, *Studies on Hysteria*. London: Hogarth Press, 1955.

Freud, S. *The Standard Edition of the Complete Psychological Works of Sigmund Freud*, vol. XVI. London: Hogarth Press, 1963.

Gould, S. J. This view of life. *Natural History*, June 1986.

Groopman, J. Hurting all over. *The New Yorker*, November 13, 2000.

Lund, N., Bengtsson, A., and Thorberg, P. Muscle tissue oxygen pressure in primary fibromyalgia, *Scand J Rheumatol* 15:165–173, 1986.

Miller, H. C. Stress prostatitis. *Urology* 32:507–510, 1988.

Pert, C. *Molecules of Emotion*. New York: Scribner, 1997.

Pimentel, M., Chow, E. J., and Lin, H. C. Eradication of small intestinal bacterial overgrowth reduces symptoms of irritable bowel syndrome, *Am J Gastroenterol* 95:3503–3506, 2000.

Reichlin, S. Neuroendocrine-immune interactions. *N Engl J Med* 329: 1246–1253, 1993.

Rosenkranz, M. A., Busse, W. W., Johnstone, T., et al. Neural circuitry underlying the interaction between emotion and asthma symptom exacerbation. Proceedings of the National Academy of Sciences 102:13319–13324, 2005.

Sapolsky, R. M. Of mice, men, and genes. *Natural History*, May 2004.

Sarno, J. E. *Healing Back Pain*. New York: Warner Books, 1991.

Sarno, J. E. *Mind over Back Pain*. New York: William Morrow & Co., 1984.

Sarno, J. E. *The Mindbody Prescription*. New York: Warner Books, 1998.

Schrader, H., Oblienne, G., Bovim, D., et al. Natural evolution of late whiplash outside the medicolegal context. *Lancet* 347 (9010):1207–1211, 1996.

Shorter, E. *From Paralysis to Fatigue*. New York: The Free Press, 1992.

Todnem, K., and Lundemo, G. Median nerve recovery in carpal tunnel syndrome. *Muscle Nerve* 23:1555–1560, 2000.

Walters, A. Psychogenic regional pain alias hysterical pain. *Brain* 84:1–18, 1961.

TWO: A BRIEF HISTORY OF PSYCHOSOMATIC MEDICINE

Alexander, F. *Psychosomatic Medicine*. New York: W.W. Norton, 1950.

Ansbacher, H. L., and Ansbacher, R. R. (editors and annotators). *The Individual Psychology of Alfred Adler* (pp. 308, 309). New York: Basic Books, 1956.

Booth, G. C. The psychological approach in therapy of chronic arthritis. *Rheumatism* 1:48, 1939.

Broks, P. What's in a face? *Prospect*, October 2000.

Cousins, N. *Anatomy of an Illness*. New York: W.W. Norton, 1979.

Draper, G. The common denominator of disease. *Am J Med Sci* 190:545, 1935.

Freud, S. *The Standard Edition of the Complete Psychological Works of Sigmund Freud*, vols. 1–33. London: Hogarth Press, 1966.

Gay, P. *Freud: A Life for Our Time* (p. 63). New York: W.W. Norton, 1988.

Halliday, J. L. Psychological aspects of rheumatoid arthritis. *Proc R Soc Med* 35:455, 1942.

Knopf, O. Preliminary report on personality studies in 30 migraine patients. *J Nerv Ment Dis* 82 (270):400, 1935.

Kohut, H. *The Analysis of the Self*. New York: International Universities Press, 1971.

Lear, J. *Love and Its Place in Nature*. New York: Farrar, Straus, Giroux, 1990.
Sarno, J. E. *Healing Back Pain*. New York: Warner Books, 1991.
Sarno, J. E. *Mind Over Back Pain*. New York: William Morrow, 1984.
Sarno, J. E. *The Mindbody Prescription*. New York: Warner Books, 1998.
Taylor, G. J. Psychosomatic Medicine and Contemporary Psychoanalysis. Madison, Conn.: International Universities Press, 1987.
Walters, A. Psychogenic regional pain alias hysterical pain. *Brain* 84:1–18, 1961.
Wolff, H. G. Personality features and reactions of subjects with migraine. *Arch Neurol Psychiat* 37:895, 1937.

THREE: THE PSYCHOLOGY OF PSYCHOSOMATIC DISORDERS

Black, P. H., and Garbutt, L. D. Stress, inflammation and cardiovascular disease. *J Psychosom Res* 52:1–23, 2002.
Coen, S. J., and Sarno, J. E. Psychosomatic avoidance of conflict in back pain. *The J Am Acad Psychoanal* 17(3):359–376, 1989.
Freud, S. *The Standard Edition of the Complete Psychological Works of Sigmund Freud*, vols. 1–33. London: Hogarth Press, 1966.
Gay, P. *Freud: A Life for Our Time* (p. 118). New York: W.W. Norton, 1988.
Lear, J. *Love and Its Place in Nature* (pp. 46–47). New York: Farrar, Straus, Giroux, 1990.
Ornish, D., Brown, S. E., Scherwitz, L.W., et al. Can lifestyle changes reverse coronary heart disease? *Lancet* 336:129–133, 1990.
Sarno, J. E. *Healing Back Pain*. New York: Warner Books, 1991.
Sarno, J. E. *Mind Over Back Pain*. New York: William Morrow & Co., 1984.
Sarno, J. E. *The Mindbody Prescription*. New York: Warner Books, 1998.

FOUR: TREATMENT

Ansbacher, H. L., and Ansbacher, R. R. (editors and annotators). *The Individual Psychology of Alfred Adler* (p. 294). New York: Basic Books, 1956.
Davanloo, H. *Basic Principles and Techniques in Short Term Dynamic Psychotherapy*. New York: Spectrum Publications, 1978.
Feinblatt, A., and Meighan, D. Short Term Group Psychotherapy for Chronic Pain Paper.
Disorders. Poster presentation, American Psychosomatic Society, April 3, 1992.
Maunder, R., and Hunter, J. Attachment and psychosomatic medicine: Developmental contributions to stress and disease. *Psychosom Med* 63:556–567, 2001.
Presentation, American Psychosomatic Society, February 13, 1998.

Sarno, J. E. *Healing Back Pain*. New York: Warner Books, 1991.
Sarno, J. E. *Mind Over Back Pain*, New York: William Morrow & Co., 1984.
Sarno, J. E. *The Mindbody Prescription*. New York: Warner Books, 1998.

FIVE: HYPERTENSION AND THE MINDBODY CONNECTION

Alexander, C. N., Schneider, R. H., Staggers, F., et al. Trial of stress reduction for hypertension in older African Americans. II. Sex and risk subgroup analysis. *Hypertension* 28:228, 1996.

Eisenberg, D. M., Delbanco T. L., Berkey, C. S., et al. Cognitive behavioral techniques for hypertension: are they effective? *Ann Intern Med* 118:964–972, 1993.

Fauvel, J. P., M'Pio, I., Quelin, P., et al. Neither perceived job stress nor individual cardiovascular reactivity predict high blood pressure. *Hypertension* 42:1112–1116, 2003.

Holtzman, J. F., Kaihlanen, P. M., Rider, A., et al. Concomitant administration of terazosin and atenolol for the treatment of essential hypertension. *Arch Intern Med* 148:539–543, 1988.

Jorgensen, R. S., Johnson, B. T., Koloodziej, M. E., et al. Elevated blood pressure and personality: A meta-analytic review. *Psychol Bull* 120:293–320, 1996.

MacMillan, H. L., Fleming, J. E., Trocme, N., et al. Prevalence of child physical and sexual abuse in the community: Results from the Ontario Health Supplement. *JAMA* 278:131, 1997.

Mann, S. J. *Healing Hypertension: A Revolutionary New Approach*. New York: Wiley, 1999.

Mann, S. J. Severe paroxysmal hypertension (pseudopheochromocytoma): Understanding its cause and treatment. *Arch Intern Med* 159:670–674, 1999.

Mann, S. J. Neurogenic essential hypertension revisited: The case for increased clinical and research attention. *Am J Hypertens* 16:881–888, 2003.

Mann, S. J., and Delon, M.A. Case report: Improved hypertension control following disclosure of decades-old trauma. *Psychosom Med* 57:501–505, 1995.

Mann, S. J., and James, G. D. Defensiveness and hypertension. *J Psychosom Res* 45:139–148, 1998.

Mann, S. J. and Gerber, L. M. Low dose alpha/beta blockade in the treatment of essential hypertension. *Am J Hypertens* 14:553–558, 2001.

Mann, S. J., and Gerber, L. M. Psychological characteristics and responses to antihypertensive drug therapy. *J Clin Hypertens* 4:25–33, 2002.

Schneider, R. H., Alexander, C. N., Staggers, F., et al. A randomized controlled trial of stress reduction in African Americans treated for hypertension for over one year. *Am J Hypertens* 18:88–98, 2005.

Suls, J., Wan, C. K., Costa, P. T. Relationship of trait anger to blood pressure: A meta-analysis. *Health Psychol* 14:444–456, 1995.

SEVEN: A RHEUMATOLOGIST'S EXPERIENCE WITH PSYCHOSOMATIC DISORDERS

Assendelft, W. J., Morton, S. C., Yu, E. I., et al. Spinal manipulative therapy for low back pain: A meta-analysis of effectiveness relative to other therapies. *Ann Intern Med* 138:871–881, 2003.

Boden, S. D., Davis, D. O., Dina, T. S., et al. Abnormal magnetic resonance scans of the lumbar spine in asymptomatic subjects: A prospective investigation. *J Bone Joint Surg [Am]* 72:403–408, 1990.

Brosseau, L., Milne, S., Robinson, V., et al. Efficacy of TENS for treatment of chronic low back pain: A meta-analysis. *Spine* 27:596–603, 2002.

Busch, A., Schachter, C. L., Peloso, P. M., et al. Exercise for treating fibromyalgia syndrome. *Cochrane Database System Review* 3:CD003789, 2002.

Deyo, R.: Diagnostic evaluation of low back pain: Reaching a specific diagnosis is often impossible. *Arch Intern Med* 162:1444–1447, 2002.

Deyo, R.: Magnetic resonance imaging of the lumbar spine: Terrific test or tar baby? *N Engl J Med*, 3331:115–116, 1994.

Deyo, R. A., Nachemson, A., Mirza, S. K. Spinal fusion surgery—The case for restraint. *N Engl J Med* 350:722–726, 2004.

Furlan, A. D., Clark, J., Esmail, R., et al. A critical review of reviews on the treatment of chronic low back pain. *Spine* 26:E155–E162, 2001.

Gerritsen, A. A., de Krom, M. C., Struijs, M. A., et al. Conservative treatment options for carpal tunnel syndrome: A systematic review of randomized controlled trials. *J Neurol* 249:272–280, 2002.

Gibson, J. N., Woddell, G., Grant, I. C., et al. Surgery for degenerative lumbar spondylosis. *Cochrane Database System Review* 3:CD001352, 2000.

Hagan, K. B., Hilde, G., Jamtretd, G., et al. The Cochrane review of advice to stay active as a single treatment of low back pain and sciatica. *Spine* 27:1736–1741, 2002.

Jellema, P., van Tulder, M. W., van Poppel, M. L., et al. Lumbar supports for prevention and treatment of low back pain: A systematic review within the framework of the Cochrane Back Review Group. *Spine* 26:377–386, 2001.

Jensen, M. C., Brant-Zawadzki, M. N., Obuchowski, N., et al. Magnetic resonance imaging of the lumbar spine in people without back pain. *N Engl J Med* 331:69–73, 1994.

Magora, A., Schwartz, A. Relation between the low back pain syndrome and x-ray findings. Parts 1–4. *Scand J Rehabil Med* 1976, 1978, 1980; 8, 10, 12.

van der Windt, D. A., van der Heijden, G. J., van den Berg, S. G., et al. Ultrasound therapy for musculoskeletal disorders: A systematic review. *Pain* 81:257–271, 1999.

van Tulder, M. W., Malmivara, A., Esmail, R. et al. Exercise therapy for low back pain. *Spine* 25:2784–2796, 2000.

van Tulder, M. W., Ostelo, R., Vlaeyen, J. W., et al. Behavioral treatment for chronic low back pain: A systematic review within the framework of the Cochrane Back Review Group. *Spine* 26:270–281, 2001.

INDEX

Page numbers in *italics* refer to illustrations.